European Studies in Philosophy of Science

Volume 11

This new series results from the synergy of EPSA - European Philosophy of Science Association - and PSE - Philosophy of Science in a European Perspective: ESF Networking Programme (2008–2013). It continues the aims of the Springer series "The Philosophy of Science in a European Perspective" and is meant to give a new impetus to European research in the philosophy of science. The main purpose of the series is to provide a publication platform to young researchers working in Europe, who will thus be encouraged to publish in English and make their work internationally known and available. In addition, the series will host the EPSA conference proceedings, selected papers coming from workshops, edited volumes on specific issues in the philosophy of science, monographs and outstanding Ph.D. dissertations. There will be a special emphasis on philosophy of science originating from Europe. In all cases there will be a commitment to high standards of quality. The Editors will be assisted by an Editorial Board of renowned scholars, who will advise on the selection of manuscripts to be considered for publication.

More information about this series at http://www.springer.com/series/13909

Brigitte Falkenburg • Gregor Schiemann
Editors

Mechanistic Explanations in Physics and Beyond

Springer

AIPS

Editors
Brigitte Falkenburg
Technische Universität Dortmund
Dortmund, Germany

Gregor Schiemann
Bergische Universität Wuppertal
Wuppertal, Germany

ISSN 2365-4228 ISSN 2365-4236 (electronic)
European Studies in Philosophy of Science
ISBN 978-3-030-10706-2 ISBN 978-3-030-10707-9 (eBook)
https://doi.org/10.1007/978-3-030-10707-9

This Springer imprint is published by the registered company Springer Nature Switzerland AG.
The registered company address is: Gewerbestrasse 11, 6330 Cham, Switzerland

Preface

Mechanistic explanations explain how certain properties of a whole stem from the causal activities of its parts. In the practice of science, many explanatory models of complex systems and their behaviour employ this kind of explanation. Given that mechanistic explanations are widely spread in biology and neuroscience, the philosophy of biology and neuroscience took them up, giving rise to an extended philosophical debate on the structure and scope of mechanistic explanations. With this book, we want to broaden the scope of the discussion, going back to the roots of mechanistic explanations in physics.

The book emerged from the 2016 conference *Mechanistic Explanations, Computability and Complex Systems* of the *Académie Internationale de Philosophie des Sciences (AIPS)*, which took place at the Technische Universität Dortmund, October 28–30, 2016. The *AIPS*, the *German Research Foundation (DFG)*, and the *Society of the Friends of the Technische Universität Dortmund* generously supported the conference. Without the endorsement of the president of the *AIPS*, Gerhard Heinzmann, it would not have been possible to organize the conference and to edit this book. The book chapters emerged from the talks given at the conference, except Meinard Kuhlmann's whom we invited afterwards to contribute to the volume. In addition, we would like to thank the editors of the European Studies in Philosophy of Science, Dennis Dieks, Maria Carla Galavotti, and Wenceslao J. Gonzalez, for the possibility of publishing the book in this series. We wish to express our gratitude to Stuart Glennan for his valuable comments on the contributions to this volume.

Dortmund, Germany — Brigitte Falkenburg

Wuppertal, Germany — Gregor Schiemann

January 2018

Contents

Chapter 1
Introduction

Brigitte Falkenburg and Gregor Schiemann

1.1 Relation to New Mechanism

Biological and neuroscientific phenomena are in the fore of works on new mechanism – a dominance that may have emerged for historical, contingent reasons. New mechanism justifiably criticized the so-called covering law model of explanation as it overly relies on physical theories and is applicable to biological phenomena only to a limited extent, at best. According to the covering law model, a phenomenon is considered explained when its theoretical representation is deductively derived using a law of nature. However, whether there are any biological laws at all is a problematic issue. Despite its focus on physical theories, the influential covering law model was also subject to criticism as it did not do justice to the explanations and models of physical practices. The notion of natural law remained in need of clarification; commonly occurring statistical explanations were not strictly necessary; in the formal representation cause and effect of phenomena were interexchangeable – only to mention a few well-known issues with the use of the covering law model in physics.

As of yet, new mechanism impulses have been only partially absorbed into the philosophy of physics. Examples of an absorbed integration of new mechanism in the scientific theory of physics are: Illari and Williamson (2012), Kuhlmann and Glennan (2014) and Kuhlmann (2017). Even more astounding when one considers that throughout history, mechanistic explanations have been closely bound to physics, and that contemporary physics is rife with mechanistic explanations, often

B. Falkenburg (✉)
Technische Universität Dortmund, Dortmund, Germany
e-mail: brigitte.falkenburg@tu-dortmund.de

G. Schiemann
Bergische Universität Wuppertal, Wuppertal, Germany
e-mail: schiemann@uni-wuppertal.de

B. Falkenburg, G. Schiemann (eds.), *Mechanistic Explanations in Physics and Beyond*, European Studies in Philosophy of Science 11,
https://doi.org/10.1007/978-3-030-10707-9_1

explicitly using the term "mechanism" (see the examples given in Falkenburg's contribution. In the following, all citations without year refer to the contributions of this volume). New mechanism offers a conceptual instrumentation for a better understanding of these points of contact. On the other hand, an increased reference to physical phenomena within the mechanistic explanation discussion, to which this volume would like to contribute, will provide a new historic and systematic gain of knowledge regarding new mechanism. Thus, for example, reflecting on new mechanism's historical origins can be beneficial in clarifying its contemporary conceptual issues (for a paradigmatic example see Glennan 1992). Furthermore, as explanations in terms of levels, mechanistic explanations draw from the underlying physical conditions and processes of complex phenomena. The discussion surrounding the scope of mechanism can therefore not succeed without taking the physical foundations into account. Finally, the comprehensive claim of validity of new mechanism requires the most all-encompassing inclusion of physics.

Increasingly recognizing physical phenomena as components of mechanistic explanations triggers questions concerning the general structure of mechanistic explanations; concerning the conditions under which other sciences explain complex phenomena by drawing on physical conditions and processes; as well as questions concerning the possibilities of mechanism of computation, which are closely related to mathematical theories.

1.2 Structure and Scope of the Book

We divided the book into three parts. *Part I* is made up by four chapters that compare the traditional mechanical explanations of early modern science and philosophy to the current accounts of mechanistic explanations. The three chapters of *Part II* deal with general questions concerning the relations between mechanisms, causality, and the multi-level structure of complex systems in physics and beyond. Finally, *Part III* collects three studies that investigate the scope of mechanistic explanations in different sciences, making the bridge from physics to economics, complexity and computation theory.

A general topic of the book is the question of how current mechanistic explanations in physics and beyond relate to the roots of mechanical explanations in early modern science and philosophy. Are current mechanistic explanations legitimate successors of their early modern precursors or not? To what extent does it make sense to generalize the traditional mechanical explanations? The four chapters of *Part I* investigate these questions going back to traditional accounts of mechanical explanations, whereas the chapters of *Part III* provide case studies from the practice of science in order to shed light on the scope of mechanistic explanations.

In addition, the contributions to the book address three groups of crucial metaphysical questions. (i) What is the ontological and/or epistemological import of mechanistic explanations, and what is their methodological significance within the practice of science? (ii) How do the causal aspects of mechanisms relate to the laws of physics? (iii) How does theory reduction relate to ontological reduction, to what extent can mechanisms explain the emergence of higher-level phenomena in nature? Here, the compositional complexity of mechanisms comes into play, and in particular, the one-level or multi-level structure of mechanisms. The three chapters of *Part II* focus on these metaphysical problems, which however also are important side issues of the contributions to *Parts I* and *III*.

Obviously, the scope of mechanistic explanations crucially depends on the respective definition of a mechanism and its interpretation. The concepts of mechanisms underlying the book contributions, however, are not uniform. Almost all book chapters rely on the seminal definitions of Glennan (1996, 2002) and Machamer et al. (2000). However, they interpret the metaphysical implications of these definitions in different ways. The ways in which they relate them to their traditional precursors neither are uniform. The differences begin with the historical background sketched in the chapters of *Part I*.

1.2.1 Part I: Mechanisms in History and Today

The four chapters of *Part I* focus on the tradition of mechanistic explanations in physics. They substantially differ in the underlying traditional concepts of a mechanism as well as in the systematic aspects of mechanistic explanations they study, complementing each other nicely regarding the metaphysical and methodological aspects of mechanistic explanations. – In *Mechanisms, Then and Now: From Metaphysics to Practice*, Stathis Psillos and Stavros Ioannidis base their discussion on Descartes' mechanistic model of matter. In contradistinction to it, they consider Newton's theory of gravitation as critical of mechanism. On these grounds, they discuss the distinction of mechanism as a metaphysical thesis and as a methodological principle of scientific theories, concluding with a plea for methodological mechanism. – For Gregor Schiemann, the traditional account of a mechanism relies on the Post-Cartesian concepts of matter in motion as described by mechanics. His chapter *Old and New Mechanistic Ontologies* compares monistic and dualistic ontologies of matter and force, as represented by Huygen's and Newton's theories. Against this metaphysical background, he interprets Glennan's account of a mechanism as monistic, but the one of Machamer, Darden and Craver as dualistic. – In *Mechanisms, Explanation and Understanding in Physics*, Dennis Dieks emphasizes that mathematization is much more typical of the explanatory tradition of physics than mechanization. His starting point is a concept of mechanism (following Psillos 2011), according to which (spatial) decomposition characterizes a mechanism. He argues that classical physics does not commit to mechanistic explanations, given that there are alternatives to a Newtonian perspective on classical mechanics and to the mechanical models of Maxwell's electrodynamics. In addition, he shows that quantum mechanics is at odds with his account of a mechanism. – In *Mechanistic Explanations Generalized: How Far Can We Go?* Brigitte Falkenburg takes the seventeenth century concept of a machine as underlying. Her chapter investigates the methodological and ontological continuity of mechanistic explanations from early modern science to current scientific practice, focusing on their generalizations in physics. She discusses how they fit in with a generalized mechanistic methodology of the "dissecting" sciences, on the one hand, and philosophical generalizations of the concept of a mechanism, on the other hand.

1.2.2 Part II: Mechanisms, Causality, and Multilevel Systems

The chapters collected in *Part II* tackle systematic questions concerning mechanistic explanations. Their common topic is the role of causality in mechanistic explanations and the way in which it relates to the distinction between higher-level and lower-level mechanisms. – Michel Ghins proposes in *Mechanist Explanation: An Extension and Defence* an enlarged version of the mechanistic account of explanation, advocated by Wesley Salmon and Phil Dowe. This avoids a problematic dichotomy between mechanistic causality and fundamental causality. Furthermore, its scope of application comprises explanations of the global behaviour of complex systems, not only in physics but also in all other scientific fields. – In *Multilevel Reality, Mechanistic Explanations, and Intertheoretic Reduction*, Marco Buzzoni counters the view that questions regarding the nature of *interlevel* explanations may be addressed separately from the *intertheoretic* reduction issue. Under the assumption that mechanistic explanations depend on perspectives, which scientists explicitly or implicitly adopt, interlevel explanations and intertheoretic reductions become connected. Buzzoni introduces an ideal distinction between weak and strong relations of perspectives in order to clarify the connection. Weak relations are combined with compatible theories, strong relations with incompatible theories. Examples taken from physics, biology and cancer research demonstrate that successfully interconnecting multiple perspectives makes the development of new perspectives necessary. – With his *A Methodological Interpretation of Mechanistic Explanation,* Hans Lenk relates several current causal and mechanistic explanation approaches to his scheme-interpretationism. This methodological approach presented in several publications (e.g. Lenk 2003) comprises the philosophy of knowledge and of action. Lenk's view that grasping cognition und action depend on interpretation exhibits partial similarity to Buzzonis perspectivalism. More generally speaking, several beneficial relationships exist between methodological scheme-interpretationism and new mechanism.

1.2.3 Part III: From Physics to Complexity and Computation

The three chapters of *Part III* investigate the scope of mechanistic explanations in the theory of complex systems, economics, and computer science. One of them is critical of the new mechanical philosophy. The other two demonstrate physics-based applications of mechanistic explanations beyond physics.– Jan Faye's contribution *Causal Mechanisms, Complexity, and the Environment* suggests abandoning the vertical top-down and bottom-up models of mechanisms in favour of a horizontal perspective on the interaction of complex systems with their environment, in order to avoid the dilemma of either downward causation or epiphenomenalism in mechanistic explanations of emergent phenomena. According to this horizontal perspective, the properties of a complex system such as the behaviour of a flock of starlings are due to external properties of the stars. – In *Crossing Boundaries: Why Physics Can Help Understand Economics*, Meinard Kuhlmann shows how the mechanisms of condensed matter physics apply to economics due to formal analogies between the macro-behaviour of social collectives and complex many-particle systems, which gave rise to the discipline of econophysics. He discusses the analogy between critical phase transitions and financial market crashes and claims that the models of econophysics describe causal mechanisms. – The chapter *Realizing Computations* of Vincenzo Fano, Pierluigi Graziani, Mirko Tagliaferri, and Gino Tarozzi completes the book. They develop a philosophical notion of the implementation of a computation, according to which by a physical system realizes a computation if there is a map from the computational states to a discrete model of system. Their definition of implementation helps understanding the mechanisms of computation, given that it explains the phenomenon of computation in terms of the (modelled) parts of a physical system individuated by the implementation function they suggest.

1.3 Concluding Remarks

The question of how strict or liberal mechanistic explanations should be understood remains open and controversial. This holds for mechanisms in general as well as in particular to their application to physical phenomena. Nevertheless, it is possible to state some tendencies. In the context of the general discussion, which focuses on the practice of biological and neuroscientific research, a broad minimal definition of the mechanism has gained in importance (Craver and Tabery 2015; Glennan and Illari 2017). In view of the physical phenomena that dominated the history of the mechanism, and that still represent a great potential for applying it today, differences in defining the scope of the concept remain significant. Historical ways of understanding that go back to the early modern period (Dieks, Falkenburg, Psillos and Stavros, Schiemann) differ from opinions that refer to the practice of current physics and its applications in other disciplines (Falkenburg, Ghins, Kuhlmann). Last not least, the authors disagree on the question of whether or not quantum mechanisms exist (Dieks vs. Kuhlmann and Glennan 2014).

However the term is understood, it must do justice to the many facets of mechanistic explanations. The actually existing plurality of phenomena (Buzzoni, Dieks, Falkenburg, Faye, Psillos and Stavros) is in tense relationship with well-justified methodological claims of universality (Buzzoni, Falkenburg, Ghins, Kuhlmann).

The universality of mechanistic models works best in cases of formal analogies (Faye, Kuhlmann) or in cases of computer algorithms that can be mapped to idealized models of physical systems (Fano).

Relatively independent of the respective concept, the authors of our volume broadly agree that the potential for applying the notion of mechanistic explanations to physical phenomena is by no means exhausted, however probably limited. The concept of mechanistic explanations serves for an understanding with intuitive models (Dieks), but also opens ways for general insights of how to decompose natural phenomena into dynamic part-whole relations between complex systems and their dynamic parts (Falkenburg). The limitations of applying a mechanistic approach in the most general sense to complex phenomena show up in diverse fields of physics, and in particular at the level of the fundamental laws of quantum mechanics and relativity theories (Dieks; Glennan 1996). By discussing cases of how the concept of mechanistic explanations applies to physical phenomena, approaches can be developed which are also fruitful for the subject areas of other disciplines (Buzzoni, Fano, Ghins, Kuhlmann).

For the discussion of mechanistic explanations in the philosophy of physics, the significance of ontological questions is perhaps just as disputed as in the general discourse of the mechanism. Ontological questions relating to the existence and structure of objects play a particular role with regard to multilevel explanations, in which physical phenomena constitute the lower layer used to explain the more complex phenomena of the upper layers. Do elements of these structures exist independently of the way in which science captures them, or are they primarily due to the theoretical assumptions underlying the access to them by science? Relying on different aspects of the practice of physics, the importance of ontology is either relativized (Lenk, Psillos and Stavros), or emphasized (Falkenburg, Faye, Schiemann).

With regard to the theories of explanation developed in the philosophy of science in physics, the mechanistic approach is only able to cope with a subdomain (Dieks, Falkenburg). In spite of its explanatory limitations, however, the innovative and by no means sufficiently recognized potential of the new mechanism for the philosophy of physics should not be underestimated. After a still unfinished phase of demarcating the new mechanism against the philosophy of physics, which is due to the criticism of the Covering-Law model of explanation, we hope that in the future the mechanistic explanations of physics will once again contribute to stimulating the general discussion about mechanisms.

References

Craver, C.F., and J. G. Tabery. 2015. Mechanisms in Science. In *Stanford Encyclopedia of Philosophy*, ed. E.N. Zalta. https://plato.stanford.edu/entries/science-mechanisms [07.04.2017].

Glennan, S.S. 1992. *Mechanisms, Models, and Causation*. Ph.D. dissertation. University of Chicago, Chicago, IL.

———. 1996. Mechanisms and the Nature of Causation. *Erkenntnis* 44 (1): 49–71.

———. 2002. Rethinking Mechanistic Explanation. *Philosophy of Science* 69 (S3): 342–353.

Glennan, S.S., and P. Illari. 2017. Varieties of Mechanisms. In *The Routledge Handbook of Mechanisms and Mechanical Philosophy*, ed. S. Glennan and P. Illari, 91–103. New York: Routledge.

Illari, P., and J. Williamson. 2012. What Is a Mechanism? Thinking About Mechanisms 'Across' the Sciences. *European Journal for Philosophy of Science* 2 (1): 119–135.

Kuhlmann, M. 2017. Mechanisms in Physics (with an Enhanced Bibliography). In *The Routledge Handbook of Mechanisms and Mechanical Philosophy*, ed. S. Glennan and P. Illari, 283–295. New York: Routledge.

Kuhlmann, M., and S.S. Glennan. 2014. On the Relation Between Quantum Mechanical, and Neo-Mechanistic Ontologies and Explanatory Strategies. *European Journal for Philosophy of Science* 4: 337–359.

Lenk, H. 2003. *Grasping Reality*. Singapore: World Scientific.

Machamer, P., et al. 2000. Thinking About Mechanisms. *Philosophy of Science* 67 (1): 1–25.

Psillos, S. 2011. The Idea of Mechanism. In *Causality in the Sciences*, ed. P. McKay, F. Russo, and J. Williamson, 771–788. Oxford: Oxford University Press.

Part I
Mechanisms in History and Today

Chapter 2
Mechanisms, *Then and Now*: From Metaphysics to Practice

Stathis Psillos and Stavros Ioannidis

Abstract For many old and new mechanists, Mechanism is both a metaphysical position and a thesis about scientific methodology. In this paper we discuss the relation between the metaphysics of mechanisms and the role of mechanical explanation in the practice of science, by presenting and comparing the key tenets of Old and New Mechanism. First, by focusing on the case of gravity, we show how the metaphysics of Old Mechanism constrained scientific explanation, and discuss Newton's critique of Old Mechanism. Second, we examine the current mechanistic metaphysics, arguing that it is not warranted by the use of the concept of mechanism in scientific practice, and motivate a thin conception of mechanism (the truly minimal view), according to which mechanisms are causal pathways for a certain effect or phenomenon. Finally, we draw analogies between Newton's critique of Old Mechanism and our thesis that the metaphysical commitments of New Mechanism are not necessary in order to illuminate scientific practice.

2.1 Introduction

The mechanical worldview of the seventeenth century was both a metaphysical thesis and a scientific theory. It was a metaphysical thesis insofar as it was committed to a reductionist account of all worldly phenomena to configurations of matter in motion subject to laws. In particular, it was committed to the view that all macroscopic phenomena are caused by, and hence are accounted for, the interactions

S. Psillos (✉) · S. Ioannidis
Department of History and Philosophy of Science, University of Athens, Athens, Greece
e-mail: psillos@phs.uoa.gr; sioannidis@phs.uoa.gr

B. Falkenburg, G. Schiemann (eds.), *Mechanistic Explanations in Physics and Beyond*, European Studies in Philosophy of Science 11,
https://doi.org/10.1007/978-3-030-10707-9_2

of invisible microscopic material corpuscles. Margaret Wilson captured this view succinctly:

> The mechanism characteristic of the new science of the seventeenth century may be briefly characterised as follows: Mechanists held that all macroscopic bodily phenomena result from the motions and impacts of submicroscopic particles, or corpuscles, each of which can be fully characterised in terms of a strictly limited range of (primary) properties: size, shape, motion and, perhaps, solidity and impenetrability (1999, xiii–xiv).

But this metaphysical thesis did, at the same time, license a scientific theory of the world, viz., a certain conception of scientific explanation and of theory-construction. To offer a scientific explanation of a worldly phenomenon X was to provide a configuration Y of matter in motion, subject to laws, such that Y could cause X. A mechanical explanation then was (a species of) *causal explanation*: to explain that Y causes X was tantamount to constructing a mechanical model of how Y brings about X. The model was mechanical insofar as it was based on resources licensed by the metaphysical worldview, viz., action of particles by contact in virtue of their primary qualities and subject to laws of motion.[1]

Nearly four centuries later, the mechanical worldview has become prominent again within philosophy of science. It's become known as 'the New Mechanical Philosophy' and has similar aspirations as the old one. New Mechanism, as Stuart Glennan puts it,

> says of nature that most or all the phenomena found in nature depend on mechanisms—collections of entities whose activities and interactions, suitably organized, are responsible for these phenomena. It says of science that its chief business is the construction of models that describe, predict, and explain these mechanism-dependent phenomena (2017, 1).

So, New Mechanism too is both a view about science *and* about the metaphysics of nature. And yet, in New Mechanism the primary focus has been on scientific practice, and in particular on the use of mechanisms in discovery, reasoning and representation (cf. Glennan 2017, 12). The focus on the metaphysics of mechanisms has emerged as an attempt to draw conclusions about the ontic signature of the world starting from the concept of mechanism as it is used in the sciences. According to Glennan, as the research into the use of mechanism in science developed, "it has been clear to many participants in the discussion that metaphysical questions are unavoidable" (2017, 12). It is fair to say that New Mechanism aims to ground the metaphysics of mechanisms on the practice of mechanical explanation in the sciences.

The chief aim of this paper is to discuss the relation between the metaphysics of mechanisms and the role of mechanical explanation in the practice of science. It will do that by presenting and comparing the key tenets of Old and New Mechanism. Section 2.2 will be devoted to the seventeenth century Mechanism. It will present the basic contours of the mechanistic metaphysics and show how it constrained scientific explanation, focusing on the case of gravity. In this section, we will also

---

[1] For the purposes of this paper, we ignore issues of mind-body causation and we focus on body-body causation. We also ignore divisions among mechanists concerning the nature of corpuscles, the existence of vacuum etc.

discuss Isaac Newton's critique of mechanism and highlight the significance of his key thought, viz., that causal explanation should identify the causes and the laws that govern their action, irrespective of whether or not these causes can be taken to satisfy certain (mostly metaphysically driven) constraints, such as being modelled in terms of configurations of matter in motion. Section 2.3 will focus on the current mechanistic metaphysics and show that it is not warranted by the use of the concept of mechanism in scientific practice. It will show that the currently popular minimal general characterisation of a mechanism is still metaphysically inflated in various ways and will motivate a thin conception of mechanism, which is not committed to any views about the ontological signature of mechanism. This thin conception—what we call 'truly minimal mechanism'—takes it that mechanisms are causal pathways for a certain effect or phenomenon. Finally, in Sect. 2.4 we will draw analogies between Newton's critique of Old Mechanism and our critique of New Mechanism. Briefly put, the point will be that causal explanation in the sciences is legitimate even if we bracket the issue of "what mechanisms or causes are as things in the world" (Glennan 2017, 12); or the issue of what activities are and how they are related to powers and the like. The metaphysical commitments of New Mechanism are not necessary in order to illuminate scientific practice.

2.2 Old Mechanism: From Metaphysics to Practice

A rather typical example of the interplay between the metaphysical worldview and the scientific conception of the world in the seventeenth century was the attempted mechanical explanation of gravity.

2.2.1 Mechanical Models of Gravity

Let us start with René Descartes. The central aim of the 3rd and 4th part of Descartes's *Principia Philosophiae*, published in 1644, was the construction of an account of natural phenomena. In Cartesian physics, the possible empirical models of the world are restricted from above by a priori principles which capture the fundamental laws of motion and from below by experience. Between these two levels there are various theoretical hypotheses, which constitute the proper empirical subject-matter of science. These are mechanical hypotheses; they refer to configurations of matter in motion. As Descartes explains in (III, 46) of the *Principia*, since it is a priori possible that there are countless configurations of matter in motion that can underlie the various natural phenomena, "unaided reason" is not able to figure out the right configuration of matter in motion. Mechanical hypotheses are necessary but experience should be appealed to, in order to pick out the correct one:

> [W]e are now at liberty to assume anything we please [about the mechanical configuration], provided that everything we shall deduce from it is {entirely} in conformity with experience (III, 46; 1982, 106).

These mechanical hypotheses aim to capture the putative causes of the phenomena under investigation (III, 47). Hence, they are explanatory of the phenomena. Causal explanation—that is, mechanical explanation—proceeds via decomposition. It is a commitment of the mechanical philosophy that the behaviour of observable bodies should be accounted for on the basis of the interactions among their constituent parts and particles; hence, on the basis of unobservable entities. In (IV, 201; 1982, 283), Descartes states that sensible bodies are composed of insensible particles. But to get to know these particles and their properties a *bridge principle* is necessary; that is, a principle that connects the micro-constituents with the macro-bodies. According to this principle, the properties of the minute particles should be modelled on the properties of macro-bodies. Here is how Descartes put it:

> Nor do I think that anyone who is using his reason will be prepared to deny that it is far better to judge of things which occur in tiny bodies (which escape our senses solely because of their smallness) on the model of those which our senses perceive occurring in large bodies, than it is to devise I know not what new things, having no similarity with those things which are observed, in order to give an account of those things [in tiny bodies]. {E.g., prime matter, substantial forms, and all that great array of qualities which many are accustomed to assuming; each of which is more difficult to know than the things men claim to explain by their means} (IV, 201; 1982, 284).

In this passage Descartes does two things. On the one hand, he advances a *continuity thesis*: it is simpler and consonant with what our senses reveal to us to assume that the properties of micro-objects are the same as the properties of macro-objects. This continuity thesis is primarily methodological. It licenses certain kinds of explanations: those that endow matter in general, and hence the unobservable parts of matter, with the properties of the perceived bits of matter. It therefore licenses as explanatory certain kinds of unobservable configurations of matter; viz., those that resemble perceived configurations of matter. On the other hand, however, Descartes circumscribes mechanical explanation by noting *what it excludes*; that is by specifying what does not count as a proper scientific explanation. He's explicit that the Aristotelian-scholastic metaphysics of substantial forms and powerful qualities is precisely what is abandoned as explanatory by the mechanical philosophy[2].

All this was followed in the investigation of the mechanism of gravity and the (in)famous vortex hypothesis according to which the planets are carried by vortices around the sun. A vortex is a specific configuration of matter in motion—matter revolving around a centre. The underlying mechanism of the planetary system then is a system of vortices:

> [T]he matter of the heaven, in which the Planets are situated, unceasingly revolves, like a vortex having the Sun as its center, and [...] those of its parts which are close to the Sun move more quickly than those further away; and [...] all the Planets (among which we {shall from now on} include the Earth) always remain suspended among the same parts of this heavenly matter (III, 30; 1982, 196).

[2] In (IV, 204; 1982, 286) Descartes accepts that scientific explanation does not require the truth of the claims about the microconstituents of things. In the next paragraph, however, he argues that his explanations have 'moral certainty' (IV, 205; 1982, 286–7).

The very idea of this kind of configuration is suggested by experience, and by means of the bridge principle it is transferred to the subtle matter of the heavens. Hence, invisibility doesn't matter. The bridge principle transfers the explanatory mechanism from visible bodies to invisible bodies. More specifically, the specific continuity thesis used is the motion of "some straws {or other light bodies}… floating in the eddy of a river where the water doubles back on itself and forms a vortex as it swirls" (op.cit.). In this kind of motion we can see that the vortex carries the straws "along and makes them move in circles with it" (op.cit.). We also see that

> some of these straws rotate about their own centers, and that those which are closer to the center of the vortex which contains them complete their circle more rapidly than those which are further away from it (op.cit.).

More importantly for the explanation of gravity, we see that

> although these whirlpools always attempt a circular motion, they practically never describe perfect circles, but sometimes become too great in width or in length (op.cit.).

Given the continuity thesis, we can transfer this mechanical model to the motion of the planets and "imagine that all the same things happen to the Planets; and this is all we need to explain all their remaining phenomena" (op.cit.). Notably, the continuity thesis offers a heuristic for discovering plausible mechanical explanations.

Christiaan Huygens (1690) came to doubt the vortex theory "which formerly appeared very likely" to him (1997, 32). He didn't thereby abandon the key tenet of mechanical philosophy. For Huygens too the causal explanation of a natural phenomenon had to be mechanical. He said referring to Descartes:

> Mr Descartes has recognized, better than those that preceded him, that nothing will be ever understood in physics except what can be made to depend on principles that do not exceed the reach of our spirit, such as those that depend on bodies, deprived of qualities, and their motions (1997, 1–2).

Huygens posited a fluid matter that consists of very small parts in rapid motion in all directions and which fills the spherical space that includes all heavenly bodies. Since there is no empty space, this fluid matter is more easily moved in circular motion around the centre, but not all parts of it move in the same direction. As Huygens put it "it is not difficult now to explain how gravity is produced by this motion" (1997, 16). When the parts of the fluid matter encounter some bigger bodies, like the planets: "these bodies [the planets] will necessarily be pushed towards the center of motion, since they do not follow the rapid motion of the aforementioned matter" (op.cit.). And he added:

> This then is in all likelihood what the gravity of bodies truly consists of: we can say that this is the endeavor that causes the fluid matter, which turns circularly around the center of the Earth in all directions, to move away from the center and to push in its place bodies that do not follow this motion (op.cit).

In fact, Huygens devised an experiment with bits of beeswax to show how this movement towards the centre can take place.

Newton of course challenged all this, and along the line, the very idea that causal explanation should be *mechanical*. But before we take a look at his reasons and their importance for the very idea of scientific explanation, we should not fail to see the broader metaphysical grounding of the mechanical project. For, as we noted, in the seventeenth century Mechanism offered the metaphysical foundation of science.

2.2.2 Mechanical vs Non-mechanical Explanation

The contours of this endeavour are well-known. Matter and motion are the 'ultimate constituents' of nature; or, as Robert Boyle (1991, 20) put it, the "two grand and most catholic principles of bodies". Hence, all there is in nature (but clearly not the Cartesian minds) is determined (caused) by the mechanical affections of bodies and the mechanical laws. Here is Boyle again:

> [T]he universe being once framed by God, and the laws of motion being settled and all upheld by his incessant concourse and general providence, the phenomena of the world thus constituted are physically produced by the mechanical affections of the parts of matter, and what they operate upon one another according to mechanical laws (1991, 139).

The Boylean conception, pretty much like the Cartesian, took it that the new mechanical approach acquired content by excluding the then dominant account of explanation in terms of "real qualities": the scholastics "attribute to them a nature distinct from the modification of the matter they belong to, and in some cases separable from all matter whatsoever" (1991, 15–16). Explanation based on real qualities, which are distinct (and separable) from matter, is not a genuine explanation. They are posited without "searching into the nature of particular qualities and their effects" (1991, 16). They offer *sui generis* explanations: why does snow dazzle the eyes? Because of "a quality of whiteness that is in it, which makes all very white bodies produce the same effect" (1991, 16). But what is whiteness? No further story about its nature is offered, but just that it's a "real entity" inhering in the substance: why do white objects produce this effect rather than that? Because it is in their nature to act thus.

Descartes made this point too when, in his *Le Monde*, he challenged the scholastic rivals to explain how fire burns wood, if not by the incessant and rapid motion of its minute parts. In his characteristic upfrontness, Descartes contrasted two ways to explain how fire burns wood. The first is the Aristotelian way, according to which "the 'form' of fire, the 'quality' of heat, and the 'action' of burning" are "very different things in the wood" (Descartes 2004, 6). The other is his own mechanistic way: when the fire burns wood,

> it moves the small parts of the wood, separating them from one another, thereby transforming the finer parts into fire, air, and smoke, and leaving the larger parts as ashes (2004, 6).

This causal explanation, based as it is on matter in motion, is preferable precisely because it is explanatory of the burning; in contrast, the Aristotelian is not, precisely because it does not make clear the mechanism by which the fire consumes the wood:

> [Y]ou can posit 'fire' and 'heat' in the wood and make it burn as much as you please: but if you do not suppose in addition that some of its parts move or are detached from their neighbours then I cannot imagine that it would undergo any alteration or change (2004, 6).

To the then dominant account of real qualities the new mechanical metaphysics juxtaposed a different view of qualities. For something to be a quality it should be determined by the mechanical affections of matter, that is, by "virtue of the motion, size, figure, and contrivance, of their own parts" (Boyle 1991, 17). Hence, there can be no change in qualities unless there is a change in mechanical affections. Though "catholic or universal matter" is common to all bodies (being, as Boyle (1991, 18) put it, "a substance extended, divisible, and impenetrable"), it is diversified by motion, which is regulated by laws.

The key point then is that the mechanical account of nature is both a metaphysical grounding of science and a way to do science: offering mechanical explanations of the phenomena. It covers everything, from the very small to the very large. Here is Boyle again:

> For both the mechanical affections of matter are to be found, and the laws of motion take place, not only in the great masses and the middle-sized lumps, but in the smallest fragments of matter; and a lesser portion of it, being as well a body as a greater, must, as necessarily as it, have its determinate bulk and figure (1991, 143).

The metaphysical grounding of mechanical explanation renders it a distinct kind of explanation, which separates it sharply from rival accounts. Concomitantly, it becomes very clear what counts as a non-mechanical alternative. An explanation couched in terms of "nature, substantial forms, real qualities, and the like" is "unmechanical" (1991, 142). But a *sui generis* chemical account of nature is unmechanical too. As Boyle put it:

> [T]hough chemical explications be sometimes the most obvious and ready, yet they are not the most fundamental and satisfactory: for the chemical ingredient itself, whether sulphur or any other, must owe its nature and other qualities to the union of insensible particles in a convenient size, shape, motion or rest, and contexture, all which are but mechanical affections of convening corpuscles (1991, 147).

The opposition to both of these non-mechanical accounts is weaved around a certain metaphysical account of the world as fundamentally mechanical and a reductive-decompositional account of scientific explanation itself.

2.2.3 Newton Against Mechanism

When Newton offered a non-mechanical account of gravity, he primarily challenged the idea that legitimate scientific explanation ought to be mechanistic. There is a sense in which Newton prioritised explanation by unification under laws and not by mechanisms. This is seen in the Preface to the second (1713) edition of the *Principia*, authored by Roger Cotes under the supervision of Newton. In this preface, Cotes presents Newton's method as a midway (a *via media*) between Aristotelianism and Mechanism. To be sure, mechanical explanations were an improvement over the Scholastic explanations because they relied on demonstrations on the basis of laws. Still, taking "the foundation of their speculations from hypotheses", the mechanists are "merely putting together a romance [i.e. fiction], elegant perhaps and charming, but nevertheless a romance" (Newton 2004, 43).

Thus put, the point sounds epistemic; it concerns the increased risk involved in hypothesising a mechanism which is supposed to underpin, and hence to causally explain, a certain phenomenon. Cotes adds:

> But when they [the mechanists] take the liberty of imagining that the unknown shapes and sizes of the particles are whatever they please, and of assuming their uncertain positions and motions, and even further of feigning certain occult fluids that permeate the pores of bodies very freely, since they are endowed with an omnipotent subtlety and are acted on by occult motions: when they do this, they are drifting off into dreams, ignoring the true constitution of things, which is obviously to be sought in vain from false conjectures, when it can scarcely be found out even by the most certain observations (Newton 2004, 43).

Still, it's fair to say that Newton's *via media* was based on a different understanding of scientific explanation: it should look for causes—hence, scientific explanation should be causal—but the sought after causes need not be mechanical. Newton's way, Cotes says, is to "hold that the causes of all things are to be derived from the simplest possible principles", but unlike the mechanists's way, it "assume(s) nothing as a principle that has not yet been thoroughly proved from phenomena". The "explication of the system of the world most successfully deduced from the theory of gravity" is the "most illustrious" example of Newton's way (2004, 32).

Newton emphatically denied feigning any hypotheses about the cause of gravity. For him,

> it is enough that gravity really exists and acts according to the laws that we have set forth and is sufficient to explain all the motions of the heavenly bodies and of our sea (2004, 92).

Gravity according to Newton is a non-mechanical force since it

> operates not according to the quantity of the surfaces of the particles upon which it acts (as mechanical causes used to do), but according to the quantity of the solid matter which they contain, and propagates its virtue on all sides to immense distances, decreasing always as the inverse square of the distances (op.cit.).

He added that the very motion of the comets makes it plausible to think that the regular elliptical motion of the planets (as well as of their satellites) cannot "have their origin in mechanical causes" (2004, 90).

In his already mentioned *Discourse on the Cause of Gravity* (1690), Huygens expressed his dissatisfaction with Newton's failure to offer a *mechanical* explanation of the cause of gravitational attraction. Favouring his own explanation of gravity in terms of the centrifugal force of the subtle and rapidly moving matter that fills the space around the Earth and the other planets, Huygens noted that Newton's theory supposes that gravity is "an inherent quality of corporeal matter". "But", he immediately added, such a hypothesis "would distance us a great deal from mathematical or mechanical principles" (1997, 35).

Yet Huygens had no difficulty in granting that Newton's law of gravity was essentially correct when it comes to accounting for the planetary system. As he put it:

> I have nothing against *Vis Centripeta*, as Mr. Newton calls it, which causes the planets to weigh (or gravitate) toward the Sun, and the Moon toward the Earth, but here I remain in agreement without difficulty because not only do we know through experience that there is such a manner of attraction or impulse in nature, but also that it is explained by the laws of motion, as we have seen in what I wrote above on gravity (1997, 31).

Explaining the fact that gravity depends on the masses and diminishes with distance "in inverse proportion to the squares of the distances from the centre" (1997, 37), were, for Huygens, clear achievements of Newton's theory despite the fact that the mechanical cause of gravity remained unidentified.

Commitment to mechanical explanation was honoured by Gottfried Wilhelm Leibniz too. In a piece titled "Against Barbaric Physics: Toward a Philosophy of What There Actually Is and Against the Revival of the Qualities of the Scholastics and Chimerical Intelligences" (written between 1710 and 1716), he defended the mechanical view by arguing that corporeal forces should be grounded mechanically when it comes to their application to the natural world. Leibniz was very clear that though he allowed "magnetic, elastic, and other sorts of forces", they are permissible "only insofar as we understand that they are not primitive or incapable of being explained, but arise from motions and shapes" (Leibniz 1989, 313). So, forces are necessary, but a condition for their applicability to the natural world is that they are seen as "arising from motions and shapes". What he took it to be "barbarism in physics" was to posit *sui generis*, that is non-mechanically grounded, "attractive and repulsive" forces that act at a distance (1989, 314–315). Newton's gravity was supposed to be such a barbaric force!

In a letter he sent to Nicolaas Hartsoeker (Hanover, 10 February 1711), Leibniz makes it clear that the proper scientific explanation should be mechanical. It is not enough for scientific explanation to identify the law by means of which a certain force acts; what is also required is the specification of the mechanism by means of which it acts. The mechanism is, clearly, on top of the law and given independently of it. Without the mechanism the power is "an unreasonable occult quality". He says:

> Thus the ancients and the moderns, who own that gravity is an occult quality, are in the right, if they mean by it that there is a certain mechanism unknown to them, whereby all bodies tend towards the center of the earth. But if they mean that the thing is performed without any mechanism by a simple primitive quality, or by a law of God, who produces that effect without using any intelligible means, it is an unreasonable occult quality, and so very occult, that it is impossible it should ever be clear, though an angel, or God himself, should undertake to explain it (Newton 2004, 112).

Newton couldn't disagree more. In an unsent letter written circa May 1712 to the editor of the *Memoirs of Literature*, Newton referred explicitly to Leibniz's letter to Hartsoeker, and stressed that it is not necessary for the introduction of a power—such as gravity—to specify anything other than the law it obeys; no extra requirements should be imposed, and in particular no requirement for a mechanical grounding:

> And therefore if any man should say that bodies attract one another by a power whose cause is unknown to us, or by a power seated in the frame of nature by the will of God, or by a power seated in a substance in which bodies move and float without resistance and which has therefore no vis inertiae but acts by other laws than those that are mechanical: I know not why he should be said to introduce miracles and occult qualities and fictions into the world. For Mr. Leibniz himself will scarce say that thinking is mechanical as it must be if to explain it otherwise be to make a miracle, an occult quality, and a fiction (Newton 2004, 116).

Note well Newton's point. The fact that an explanation does not conform to a certain mechanical framework does not make it fictitious, occult or miraculous. Non-mechanical explanations are legitimate insofar as they identify the law that covers or governs a certain phenomenon. Hence, Newton promotes a methodological shift: causal explanation without mechanisms but subject to laws.

Causal explanation then need not be mechanical to be legitimate and adequate. This is Newton's key thought. In breaking with a tradition which brought under the same roof a certain metaphysical conception of the world and a certain view of scientific explanatory practice, Newton distinguished the two and laid emphasis on the explanatory practice itself, thereby freeing it from a certain metaphysical grounding.

Though this is not the end of the story of Old Mechanism,[3] Newton's key thought, we shall argue, is of relevance in the current debates over New Mechanism, to which we shall now turn our attention.

2.3 New Mechanism: From Practice to Metaphysics

It is useful to differentiate between two ways to conceptualise mechanisms in the post-1970 literature. First, mechanism has been used as a primarily metaphysical concept, mostly aiming to illuminate the metaphysics of causation. Second, mechanism has been taken to be a concept used in science, and philosophical accounts of mechanism have aimed to elucidate this concept. To be sure, some philosophical approaches to mechanism, most notably Glennan's (1996), blend these two conceptions (the metaphysical one and the concept-in-use). However, it's fair to say that there are two quite distinct points of origin of the recent philosophical accounts of mechanism: the first starts from metaphysics (as was the case for Descartes and other old mechanists), the second from scientific practice. Using this distinction between mechanism as a primarily metaphysical concept and as a concept-in-use in science, we can differentiate between two kinds of approaches to the metaphysics of mechanisms.

[3] For more on the development of Old Mechanism, see Psillos (2011).

On the first approach, the aim is to show what the connection is between mechanism qua a metaphysical category and other central metaphysical concepts, notably causation. In the context of the metaphysics of causation, 'mechanistic' accounts are theories about the link between cause and effect. Such theories are meant to be anti-Humean in that they view causation as a productive relation, i.e. the cause somehow brings about or produces the effect. The aim of the mechanistic view of causation is to illuminate the productive relation between the cause and the effect by positing a mechanism that connects them, and by explicating 'mechanism' in a suitable way such that causal sequences are differentiated from non-causal ones. The central thought, then, is that A causes B if and only if there is a mechanism connecting A and B.

Two kinds of views have become prominent: those that characterise the mechanism that links cause and effect in terms of the persistence, transference or possession of a conserved quantity (Mackie 1974; Salmon 1997; Dowe 2000); and those that connect a mechanistic account to causal production with a power-based one (see Harré (1970) for an early such view). Despite their differences, these views share in common the claim that mechanisms are the ontological tie that constitutes Hume's 'secret connexion'. In our (2018), we have called such mechanisms, *mechanisms-of*. Mechanisms-of are ontological items that underlie or constitute certain kind of processes, i.e. those that can be deemed causal. Since we have dealt with these accounts in some detail elsewhere (Psillos 2002; Ioannidis and Psillos 2018), we shall not discuss them further here.

On the second approach, working out a metaphysics of mechanisms is not the starting point but rather the end point of inquiry. Starting with mechanism as a concept-in-use in science, one tries first to give a general characterisation of this concept and then to derive metaphysical conclusions, i.e. conclusions about the (mechanistic) structure of the world.

This kind of bottom-up inquiry has yielded several well-known general accounts of mechanisms as well as theses about the ontic signature of a mechanistic world.

2.3.1 The Metaphysics of New Mechanism

Here are three well-known general characterisations of a mechanism in recent mechanistic literature:

> Mechanisms are entities and activities organized such that they are productive of regular changes from start or set-up to finish or termination conditions (Machamer et al. 2000, 3).

> A mechanism for a behavior is a complex system that produces that behavior by the interaction of a number of parts, where the interactions between parts can be characterized by direct, invariant, change-relating generalizations (Glennan 2002, S344).

> A mechanism is a structure performing a function in virtue of its component parts, component operations, and their organization. The orchestrated functioning of the mechanism is responsible for one or more phenomena (Bechtel and Abrahamsen 2005, 423).

The focus on mechanism as a concept-in-use is common to all three accounts; none of the three accounts can be viewed as falling under the rubric of mechanistic theories of causation. And yet, all these and similar accounts yield specific metaphysical commitments about what kind of things in the world mechanisms are. All these accounts are committed to the thesis that a general characterisation of mechanism must itself be cashed out in metaphysical terms. Hence, talk of mechanisms in science is taken to have quite direct consequences about the kind of ontology presupposed by such talk. In order to substantiate this point, let us look at the three accounts mentioned earlier in some more detail.

Peter Machamer, Lindley Darden and Carl Craver's (MDC) account is perhaps the most ontologically inflated, as it is explicitly committed to both entities *and* activities as distinct and separate ontological categories. It is thus committed to a particular view about the metaphysics of causation: causation within mechanisms is to be characterised in terms of *production*, where the productive relation is captured by the various different kinds of activities identified by science.

Glennan's case is interesting, since in his (2002) he refrains from taking mechanisms to entail a productive account of causation. Instead, within-mechanism interactions are characterised in terms of invariant, change-relating generalisations. As we will see below, however, Glennan has presently connected his account of mechanisms with a power-based understanding of causation. Hence, he is committed to causal powers as parts of the building blocks of mechanisms.

Lastly, William Bechtel & Adele Abrahamsen's account does not include a specific characterisation of what mechanistic causation amounts to at all. Here, however, as in the other two accounts, we have a series of general terms the meaning of which needs to be unpacked. So, MDC include in their accounts 'entities' and 'organisation'; Glennan includes 'complex system' and 'parts'; and Bechtel & Abrahamsen talk about 'structure', 'function', 'parts' and 'organisation'.

All these accounts suggest the further need to explain what this 'new mechanical ontology' of entities, activities, organisation of parts into wholes etc. amounts to: what, in general terms, the constituents of mechanisms are and what are their relations with more traditional metaphysical categories, such as things, properties, powers and processes.

Notably, there has been a tendency recently to offer a more minimal general characterisation of a mechanism. For example, according to Phyllis Illari & Jon Williamson:

> A mechanism for a phenomenon consists of entities and activities organized in such a way that they are responsible for the phenomenon (2012, 120).

Glennan's recent version is almost identical:

> A mechanism for a phenomenon consists of entities (or parts) whose activities and interactions are organised so as to be responsible for the phenomenon (2017, 17).

Glennan calls this account *Minimal Mechanism*. The key motivation here is for a general characterisation of mechanism broad enough to capture examples of mechanisms in different fields, from physics to the social sciences. But even in this minimal mode, mechanisms, according to Glennan, "constitute the causal structure of the world" (Glennan 2017, 18).

This minimal account of mechanism might appear to fit the bill of capturing a concept-in-use in science. On closer inspection, however, it is committed to a rather rich metaphysical account of mechanism: the minimal account is not more minimal than the metaphysically inflated accounts noted above. The reason is that both of the foregoing minimal accounts still invite questions about the ontic status of mechanisms. For example: how exactly do entities and activities differ? What is the relation between activities and interactions? How should organisation be understood? Glennan (2017, 13) explicitly talks about a "new mechanical ontology" as the upshot of the minimal account. The "minimal mechanism", he adds, "is an ontological characterization of what mechanisms are as things in the world" (2017, 19).

New Mechanism, then, aims to provide a new ontology of mechanisms. We can identify three commonly accepted key theses concerning mechanistic ontology:

(1) The world consists of mechanisms.

Thesis 1 is a typical view among mechanists: mechanisms are taken to be *things in the world*, with objective boundaries. Ours is a mechanistic world. As Glennan puts it at the end of his (2017, 240), "[t]hat is just how we have found the world to be".

(2) A mechanism consists of objects of diverse kinds and sizes structured in such a way that, in virtue of their properties and capacities, engage in a variety of different kinds of activities and interactions such that a certain phenomenon P is brought about.

Thesis 2 (or something very similar) can be taken as the common core of the general characterisation of mechanism as a concept-in-use given by new mechanists. It identifies the components of a mechanism and the relations among them. As mechanisms are things in the world (thesis 1), their components are also particular things in the world. Besides, these parts engage in activities by being "active, at least potentially" (Glennan 2017, 21). Activity is understood as a manifestation of the powers things have. Glennan is quite explicit that "Activities manifest the powers (capacities) of the entities involved in the Activity" (2017, 31). Positing powers is supposed to explain why "activities are powerful"; being powerful, activities are what "an entity does, not merely something that happens to an entity" (2017, 32). But activities are not enough. Interactions are needed too because "there is no production without interaction" (2017, 22). "The fundamental point of ontological agreement among the New Mechanists", as Glennan (2017, 21 n.6) puts it, is that entities cannot exist without activities or activities without entities. It's not hard to see that the minimal account of mechanism is taken to imply or suggest a rather substantive metaphysical conception of mechanism, which, until further notice at least, is broadly neo-Aristotelian.

(3) To explain a certain phenomenon P is to offer the mechanism that produces it.

Thesis 3 connects the previous theses with a claim about explanation (and more specifically, causal explanation): since in a mechanistic world phenomena are produced by mechanisms, scientific explanation has to identify the mechanism that produces a certain phenomenon; that is (by thesis 2) to identify the organised entities and activities that produce the phenomenon.

Despite their differences, there are important similarities between Old and New Mechanism (which justify viewing both positions as mechanistic). On the one hand, as we saw, new mechanists differ from their seventeenth century predecessors in that they do not start their analysis with a metaphysical concept of mechanism; rather, they aim at giving a general characterisation of mechanism as a central concept of scientific practice. This characterisation is non-reductive in that it is not committed to the view that mechanisms are configurations of matter in motion subject to laws (and contact action). But, on the other hand, they are committed to mechanisms being configurations of powerful entities engaged in activities and interactions. As Glennan puts it: "Mechanisms are particular and compound, made up of parts (entities) whose activities and interactions are located in particular regions of space and time" (2017, 57). Hence, New Mechanism is similar to seventeenth century Mechanism, in that it is committed to a mechanistic ontology. This ontology (theses 1 & 2 above), while not a global metaphysics in the sense of the seventeenth century, is still a thesis about the ontic signature of the world. Here is Glennan again:

> New Mechanist ontology is an ontology of compound systems. It suggests that the properties and activities of things must be explained by reference to the activities and organization of their parts (2017, 57).

Instead of resulting in a 'flat' ontology where everything there is consists in matter in motion, this new mechanical metaphysics ends up with a hierarchy of particular things—mechanisms—which may contain a diverse set of entities and activities, rather than the limited set endorsed by the corpuscularians, and whose productivity is grounded in causal powers, rather than in a few fundamental laws of motion.

But we can ask: are these ontological commitments really necessary in order to understand scientific practice? Are they licenced by the practice of science? Remember here that the primary aim of new mechanists is to give a general characterisation of mechanism as a concept-in-use. So, ideally, the general account of mechanism should capture as far as possible the extension of a concept-in-use in the various sciences. The minimal account of mechanism discussed so far, though broad enough to play this role, inflates the concept-in-use by making it amenable to a certain metaphysical description of its basic components.

Note that our claim is not that the metaphysical questions are not philosophically interesting questions to ask; they are, especially if we are interested in giving an account of the ontological structure of reality. Moreover, such a kind of project has to be informed by what science has to say about the world. If, however, our aim is to understand how a specific concept—*mechanism*—is used in scientific practice, these questions seem, at least prima facie, irrelevant, especially if a general characterisation of mechanism is possible that does not include such things.

2.3.2 Mechanism in Scientific Practice

A metaphysically deflationary view of mechanism as a concept-in-use that is broad enough to capture all examples of mechanisms that we find in science seems indeed possible. In our (2017) we have argued that there is an even more minimal account of mechanism than those offered by Illari & Williamson and Glennan, which captures this concept-in-use. This is achieved by dropping the reference to activities and interactions and by understanding mechanism as the causal pathway of a certain phenomenon, described in the language of theory. According to this account that we call *Truly Minimal Mechanism* [TMM], a mechanism in science just is a causal pathway described in theoretical language:

TMM : mechanism = causal pathway, described in theoretical language

The central idea behind TMM is that when scientists talk about a 'mechanism', what they try to capture is the way (i.e. the causal pathway) a certain result is produced. Say, for example, that we want to find out how a certain disease state is brought about. What we must look for is a specific mechanism, i.e. a causal pathway that involves various causal links between, for example, a virus and changes in properties of the organism that ultimately lead to the disease state. In pathology, such causal pathways constitute the pathogenesis of a disease, and when pathologists talk about the mechanisms of a disease, it is such pathways that they have in mind (cf. Lakhani et al. 2009; see also Gillies 2017, for a similar account of mechanisms in medicine as causal pathways).

Mechanisms then and causation are closely related: when two events are causally connected, there is a mechanism (that is, a causal pathway) that connects them and accounts for the specific way that the cause brings about the effect. Also, scientists succeed in identifying a mechanism, if they succeed in describing the relevant causal pathway in terms of the theoretical language of the particular scientific field.

In life sciences, this is a typical use of 'mechanism'. In our (2017), we have discussed extensively a particular example, the case of apoptosis, i.e. a particular causal pathway (or mechanism) of cell death. The study of apoptosis (and of mechanisms of programmed and physiological cell death in general) in fact transcends particular biological fields, and has involved cytologists, developmental biologists, pathologists, and molecular biologists among others. Because of its broad role, this case offers a nice test case of the concept of mechanism as it is used in science. Apoptosis is described in various biological disciplines as a 'mechanism' of cell death. The common concept of mechanism at play here, we have argued in (2017), is that a mechanism just is a causal pathway. TMM then is the common denominator of all uses of the concept of mechanism in biology and elsewhere.

To further see the plausibility of this view, consider the relation between the concepts of mechanism and function. In the general accounts mentioned earlier, only Bechtel & Abrahamsen explicitly refer to the behaviour of a mechanism in terms of its function. But in all accounts, a mechanism is always a mechanism *for* a phenomenon or behaviour, and this can be taken as an implicit reference to the function of the mechanism, which plays a central role in individuating the mechanism. In other words, there is no mechanism without a function, and the function determines what, among everything that happens within a complex system such as an organism, counts as a mechanism. It is an open issue among new mechanists how this commitment to function should be construed and what its consequences for the metaphysics of mechanisms are (cf. Garson 2018; Craver 2013).

However, if we insist on an account of mechanism broad enough to capture all uses of the concept in science, and given that there are scientific fields where the concept of function is not present (e.g. particle physics, solid state physics, astrophysics, cosmology), an account such as TMM seems preferable. Of course, there are contexts (for example, in molecular biology), where a mechanism is automatically a mechanism for a certain function; e.g. apoptosis is a mechanism for cell death, and it also has a homeostatic function within the organism (see our (2017) for more on this point and its relevance for TMM). But it is not clear that a mechanism of star formation, for example, has star formation as its function, unless one takes function to be what the mechanism produces, i.e. its effect. The point here is that if we want to claim that a mechanism of cell death and a mechanism of star formation are in some sense the same kind of thing (i.e. they are both mechanisms), that is if we want to give a general account of a mechanism as a concept-in-use across various scientific fields, TMM seems the most promising candidate. At the same time, TMM can be easily adapted to capture particular uses of 'mechanism' in various contexts where a specific notion of function is presupposed.

We want to resist the temptation to offer a metaphysically inflated account of the causal pathway, in terms of an explicit specification of its ontological constituents. A key reason for this is that the causal pathway should be described in the *theoretical language* of a specific scientific field, and not in some privileged ontological language or even in ontologically loaded terms. This suggests that the form of the description of the mechanism cannot be decided beforehand and in advance of how the concept of mechanism is used. What counts, each time, as a legitimate description of a causal pathway, is something that has to be decided by scientific practice. Instead of imposing various metaphysical categories as those that constitute a general legitimate description of a mechanism, it should be left to the scientists themselves to decide how best to describe mechanisms using the theoretical language they employ to understand and describe the world. TMM has the consequence that a series of questions that new mechanists have been concerned with need not concern us if our aim is to understand scientific practice.[4]

If we identify a mechanism with a causal pathway, would it be required to make a commitment about what the ontological nature of causation is? This does not seem necessary for understanding the concept-in-use. Scientific practice establishes robust causal connections, which can be used for understanding and manipulation, without necessarily being committed to a single and overarching ontic account of

[4] These questions concern, for example, the components and boundaries of mechanisms (cf. Kaiser 2018), the metaphysics of causation (cf. Matthews and Tabery 2018), and the mechanistic levels (cf. Povich and Craver 2018).

causation. Ultimately, whatever fundamental ontological theory of what causation is one might have (e.g. in terms of causal powers or regularities etc.), the identification of causal relationships is based on theory-described difference-making relations; this is what scientists look for when establishing causal relations and causal pathways. In this sense, the causal pathway by means of which a phenomenon Y is brought about by a cause X, given that X initiates a chain of events that leads to Y, is the very network of theory-described difference-making relations among the various intermediaries of X and Y. It is a further question, and one that is not needed to be answered in order to discover and use causal relations in scientific practice, what the truth-makers of these difference-making relations are. Hence, the point here is that in order to understand what a causal pathway (and hence a mechanism) as a concept-in-use is, and to identify mechanisms, we do not need a theory about the metaphysics of causation: TMM is really, ontologically speaking, a truly minimal view[5].

TMM is best seen in the context of a thesis that we call, following Joseph Henry Woodger (1929) and Robert Brandon (1984), *Methodological Mechanism* [MM] (see our (2017) for more on Woodger on MM). In his (1929), Woodger distinguished between two ways in which a certain notion can be employed: a metaphysical or ontological way and a methodological one. The latter is when a notion is used for the purposes of description "independently of its metaphysical interpretation". In this case, Woodger says, the notion "is employed methodologically, i.e. simply for the purpose of investigation" (1929, 31). The advantage of this use is that the notion can be used in a certain practice and cast light on it independently of whatever difficulties (and controversies) are raised by the intricate metaphysical debates concerning what its worldly reference is really like. According to Woodger, taking the methodological standpoint amounts to asking "the methodological mechanist what he has to say in support of his contention that the mechanical explanation is the only one which is admissible in science" (1929, 231). Hence, MM is a view about mechanistic explanation and its admissibility, and not about the blueprint of the universe.

Methodological Mechanism, we want to claim, is the view that commitment to mechanism in science is adopting a methodological postulate which licenses looking for the causal pathways for the phenomena of interest. Hence, MM licenses adopting TMM.

MM illuminates practice in a way that ontologically inflated accounts of mechanism do not. It accounts for the centrality of mechanisms in scientific discovery and explanation, since according to it discovering mechanisms (i.e. causal pathways) is the central task of science. At the same time, however, it refrains from imposing on scientific practice ontic constraints that are not licensed by it. According to MM, the mechanistic view need not be taken as something stronger than a certain

[5] There are various other questions that can be raised concerning TMM: for example, do we take pathways to be types or tokens? Here again, we defer to practice. Causal pathways, qua things in the world that produce an effect, are concrete particulars. But what is described theoretically in the language of theory is a type of causal pathway.

methodological commitment to a kind of explanation. As Brandon has put it, the question: 'what is a mechanism?'

> has no general metaphysical answer, because the business of science is the discovery of mechanisms; so we cannot delimit in any a priori manner the mechanisms of nature. [...] The best we can do is to give an open-ended answer: *a mechanism is any describable causal process* (1990, 185, emphasis added).

This can be generalised as follows: concepts such as mechanism, that are central in scientific practice, should be viewed as methodological postulates rather than as presupposing robust metaphysical commitments. But methodological postulates should be 'open-ended'; otherwise they would unnecessarily limit research. Far from being a trivial commitment, MM is flexible enough to foster searching for mechanisms, whatever the ontic signature of the world might be.

2.4 Newton Revisited

How does all this get connected to Newton's critique of the Old Mechanism? In a letter to Leibniz dated 16 October 1693, Newton challenged him to offer a mechanical explanation of "gravity along with all its laws by the action of some subtle matter" and to show "that the motion of planets and comets will not be disturbed by this matter". If this were available, Newton said, he would be "far from objecting". But no such explanation was forthcoming and Newton was happy to re-iterate his view that

> since all phenomena of the heavens and of the sea follow precisely, so far as I am aware, from nothing but gravity acting in accordance with the laws described by me; and since nature is very simple [...] all other causes are to be rejected (Newton 2004, 108–109).

Newton does not simply say that causal explanation might not be mechanical. His point is that causal explanation should be liberated from mechanism. It'd not be enough to offer a mechanical account of the cause of gravity; the laws that gravity obeys should be mechanically explicable and, as Newton repeatedly stressed, this was not forthcoming. Though causal explanation matters, it doesn't matter if it is subject to various (old) mechanical constraints.

We noted already that the new mechanical conception of nature is far from the seventeenth century conception that everything should be accounted for in terms of (configurations of) matter in motion. So it's far from us to tar New Mechanism with the same brush as Old Mechanism. For instance, the key ontology of the old mechanical picture was justified, by and large, a priori, whereas the key ontology of New Mechanism is grounded in scientific practice; in this case, it is practice that constrains metaphysics. Be that as it may, we are now going to argue that there exists a kind of Newtonian move against New Mechanism too.

What is clear from the present discussion is that, regardless of the main difference noted above, the new idea of mechanism is no less metaphysically loaded than the old one. Where the seventeenth century mechanists looked for stable arrangements of matter in motion subject to laws, the twenty-first century mechanists look for stable arrangements of powerful entities engaged in various activities and interactions. These mechanisms are supposed to be the building blocks of nature and the scientific task is to unravel them. They underpin "mechanistic explanations" which, as Glennan put it, show "how the organized activities and interactions of some set of entities cause and constitute the phenomenon to be explained" (2017, 223). Mechanistic explanation "always involves characterizing the activities and interactions of a mechanism's parts" (2017, 223). Where the seventeenth century mechanists saw 'action by contact' as a requisite for a proper mechanical explanation, new mechanists see powers and 'activities'.

Why is Newton's key thought relevant to the modern debates about mechanisms? The key thought, to repeat, was that causal explanation should identify causes and the laws that govern their action irrespective of whether or not these causes can be taken to satisfy further (mostly metaphysically driven) constraints. In other words, Newton showed that certain causal explanations of phenomena (in terms of non-mechanical forces) are both legitimate and complete insofar as they identify the right causes and are empirically grounded.

We take it that the point MM stresses, is, *mutatis mutandis*, analogous to Newton's. The point of MM is that causal explanation need not be mechanistic in the new mechanists' ontic sense, and that being couched in the way new mechanists propose, causal explanation is subjected to constraints unwarranted by scientific practice. Insofar as mechanism is a concept-in-use in science, it may well be seen referring to the causal pathway of the phenomenon to be explained, couched in the language of theories. Preserving the spirit of Newton's key thought, we might say that causal explanation is legitimate even if we bracket the issue of "what mechanisms or causes are as things in the world" (Glennan 2017, 12) or the issue of what activities are and how they are related to powers and the like. The issue then is not "an ontological characterization of what mechanisms are as things in the world" (2017, 19), but a methodological characterisation of them as causal pathways described in the language of theories.

To press the analogy a bit more, questions such as:

> If entities, activities, and the mechanisms they constitute are compounds, of what are they compounded? Where does one entity or activity or mechanism end, and when does another begin? And on what account do we decide that a collection of interacting entities is to count as a whole mechanism? (Glennan 2017, 29)

are pretty much like the questions concerning the cause of the properties of gravity that Newton thought need not be asked and answered for a scientifically legitimate conception of causal explanation.

We don't want to claim that questions such as the above are not connected to scientific practice. After all, even the question of the cause of gravity that Newton refrained from answering, was connected to scientific practice. The point, rather, we take from Newton is that answering these questions is not required for offering adequate causal explanations of the phenomena under study. Similarly, for MM, answering questions such as the above is not required in order to have legitimate mechanistic explanations. In other words, the properties of mechanism over and above those that are required by its methodological use need not be specified; nor is there an explanatory lacuna if they are not.

According to MM, the concept of mechanism as used in practice need not, and should not, be understood in a metaphysically inflated sense. Hence, new mechanists, in offering such metaphysically inflated accounts, need to show that such accounts are indeed indispensable for doing good mechanistic science.

To conclude, as Newton remained agnostic about the underlying mechanism of gravity, so MM remains agnostic about the metaphysical ground of any particular causal pathway. As in the case of gravity, *it is enough that mechanisms qua causal pathways really exist and act as they do.*

Acknowledgement An earlier version of this paper was presented at the AIPS Conference on Mechanistic Explanations, in Dortmund, 27–30 October 2016. We would like to thank the audience for comments and suggestions. Special thanks are due to Brigitte Falkenburg and Gregor Schiemann for their invitation, encouragement and comments. Many thanks to Stuart Glennan for useful comments on an earlier draft.

References

Bechtel, W., and A. Abrahamsen. 2005. Explanation: A Mechanistic Alternative. *Studies in History and Philosophy of Biological and Biomedical Sciences* 36: 421–441.

Boyle, R. 1991. *Selected Philosophical Papers of Robert Boyle*, ed. M.A. Stewart. Indianapolis/Cambridge: Hackett Publishing Company.

Brandon, R.N. 1984. Grene on Mechanism and Reductionism: More Than Just a Side Issue. *PSA: Proceedings of the Biennial Meeting of the Philosophy of Science Association* 1984 (2): 345–353.

———. 1990. *Adaptation and Environment*. Princeton: Princeton University Press.

Craver, C.F. 2013. Functions and Mechanisms: A Perspectivalist View. In *Functions: Selection and Mechanisms*, ed. P. Huneman, 133–158. Dordrecht: Springer.

Descartes, R. 1982. *Principles of Philosophy*. Trans. V.R. Miller, and R.P. Miller. Dordrecht: D. Reidel Publishing Company.

———. 2004. *René Descartes: The World and Other Writings*, ed. S. Gaukroger. Cambridge University Press.

Dowe, P. 2000. *Physical Causation*. Cambridge: Cambridge University Press.

Garson, J. 2018. Mechanisms, Phenomena, and Functions. In *The Routledge Handbook of Mechanisms and Mechanical Philosophy*, ed. S. Glennan and P.M. Illari, 104–115. New York: Routledge.

Gillies, D. 2017. Mechanisms in Medicine. *Axiomathes* 27: 621–634.

Glennan, S. 1996. Mechanisms and The Nature of Causation. *Erkenntnis* 44: 49–71.

———. 2002. Rethinking Mechanistic Explanation. *Philosophy of Science* 69: S342–S353.

———. 2017. *The New Mechanical Philosophy*. Oxford: Oxford University Press.

Harré, R. 1970. *The Principles of Scientific Thinking*. London: Macmillan.

Huygens, C. 1997 [1690]. *Discourse on the Cause of Gravity*. Trans. Karen Bailey. Mimeographed.

Illari, P.M., and J. Williamson. 2012. What is a Mechanism? Thinking about Mechanisms Across the Sciences. *European Journal of Philosophy of Science* 2: 119–135.

Ioannidis, S., and S. Psillos. 2017. In Defense of Methodological Mechanism: The Case of Apoptosis. *Axiomathes* 27: 601–619.

———. 2018. Mechanisms, Counterfactuals and Laws. In *The Routledge Handbook of Mechanisms and Mechanical Philosophy*, ed. S. Glennan and P.M. Illari, 144–156. New York: Routledge.

Kaiser, M.I. 2018. The Components and Boundaries of Mechanisms. In *The Routledge Handbook of Mechanisms and Mechanical Philosophy*, ed. S. Glennan and P.M. Illari, 116–130. New York: Routledge.

Lakhani, S., S. Dilly, and C. Finlayson. 2009. *Basic Pathology: An Introduction to the Mechanisms of Disease*. 4th ed. London: Hodder Arnold.

Leibniz, G.W. 1989. *G. W. Leibniz: Philosophical Essays*, ed. R. Ariew and D. Garber. Indianapolis/Cambridge: Hackett Publishing Company.

Machamer, P., L. Darden, and C.F. Craver. 2000. Thinking About Mechanisms. *Philosophy of Science* 67: 1–25.

Mackie, J.L. 1974. *The Cement of the Universe*. Oxford: Clarendon Press.

Matthews, L.J., and J. Tabery. 2018. Mechanisms and the metaphysics of causation. In *The Routledge Handbook of Mechanisms and Mechanical Philosophy*, ed. S. Glennan and P.M. Illari, 131–143. New York: Routledge.

Newton, I. 2004. *Philosophical Writings*, ed. A. Janiak. Cambridge: Cambridge University Press.

Povich, M., and C.F. Craver. 2018. Mechanistic Levels, Reduction and Emergence. In *The Routledge Handbook of Mechanisms and Mechanical Philosophy*, ed. S. Glennan and P.M. Illari, 185–197. New York: Routledge.

Psillos, S. 2002. *Causation and Explanation*. Montreal: Acumen & McGill-Queens University Press.

———. 2011. The Idea of Mechanism. In *Causality in the Sciences*, ed. P.M. Illari, F. Russo, and J. Williamson, 771–788. Oxford: Oxford University Press.

Salmon, W. 1997. Causality and Explanation: A Reply to Two Critiques. *Philosophy of Science* 64: 461–477.

Wilson, M.D. 1999. *Ideas and Mechanism: Essays on Early Modern Philosophy*. Princeton: Princeton University Press.

Woodger, J.H. 1929. *Biological Principles: A Critical Study*. London: Routledge & Kegan Paul Ltd.

Chapter 3
Old and New Mechanistic Ontologies

Gregor Schiemann

Abstract The concept of mechanistic philosophy dates back to the beginning of the early modern period. Among the commonalities that some of the conceptions of the main contemporary representatives share with those of the leading early modern exponents is their ontological classification: as regards their basic concepts, both contemporary and early modern versions of mechanism can be divided into monist and dualist types. Christiaan Huygens' early modern mechanistic explanation of non-material forces and Stuart S. Glennan's contemporary conception of mechanism will serve as examples of monism. As examples of dualism, I will discuss Isaac Newton's early modern mechanistic philosophy of nature and the contemporary conception of mechanism proposed by Peter Machamer, Lindley Darden, and Carl F. Craver. With the ontological commonalities are associated further characteristic features of the respective types that concern, among other things, the respective understandings of fundamental theories and evaluations of scientific practice. The ontological continuity of the types does not play any role in contemporary discussions of the history of mechanistic philosophy. On my assessment the distinction between monism and dualism remains an unsolved problem and its persistence is an indication that this distinction is a fundamental one.

3.1 Introduction

It is a matter of controversy whether one can formulate a uniform concept of mechanistic philosophy for a certain period and whether it can be traced back to the early modern era. While its leading contemporary exponents portray themselves as a

G. Schiemann (✉)
Bergische Universität Wuppertal, Wuppertal, Germany
e-mail: schiemann@uni-wuppertal.de

B. Falkenburg, G. Schiemann (eds.), *Mechanistic Explanations in Physics and Beyond*, European Studies in Philosophy of Science 11,
https://doi.org/10.1007/978-3-030-10707-9_3

joint movement,[1] early modern mechanism is a post facto reconstruction that can be undertaken in a variety of ways. In the context of early modern philosophy, mechanism can mean the orientation to physical mechanics, the use of the machine metaphor to explain natural phenomena, or the mathematization of scientific knowledge—to name only a few, not always easily distinguishable, examples.[2] Early modern mechanistic thought dealt primarily with inanimate nature, though it also extended to the phenomena of life. For the biological sciences, Daniel J. Nicholson claims that no continuity exists between the important early modern meanings of mechanism and the contemporary meanings, which deal mainly with living phenomena.[3] However, the leading contemporary exponents by no means deny that there is some common ground with early modern mechanism.[4]

My comparison between the historically widely separated ontological meanings is based on points of contact between a present-day concept of mechanism and one applied to early modern philosophy. For the present-day concept of mechanism. I refer to texts by authors who are generally recognized as its leading exponents. Their conceptions are commonly divided up into the three groups comprising (1) Peter Machamer, Lindley Darden, and Carl F. Craver (hereafter abbreviated as "MDC"), (2) Stuart Glennan, and (3) William Bechtel and Adele Abrahmson. The epistemic and methodological commonalities that exist between them have been repeatedly highlighted.[5] For early modern mechanism, I propose a broad and a narrow concept that to a sufficient extent include the conceptions that were influential at that time while exhibiting points of contact with the contemporary concept. Mechanism, broadly conceived, treats matter in motion as the first and only cause of all natural phenomena and, narrowly conceived, it postulates that the forms of motion are determined by the principles of a specific discipline, namely, mechanics. In what follows I will confine myself to the narrow concept.

Assuming the concepts of early modern and contemporary mechanism, we can draw up a rough list of their commonalities and differences. As commonalities, I would cite the search for causal explanations of phenomena that are not based on supernatural forces, the rejection of a categorical distinction between natural and technical phenomena, the quest for a unified scientific method, the close connection to scientific practice, and, finally, the correspondence that concerns me in the present text—the possibility of classifying some conceptions under the same ontologi-

---

[1] Levy and Bechtel 2016 use the term "mechanism 1.0" to refer to some of the commonalities. Illari and Williamson 2012, Craver and Tabery 2015, Glennan 2016, and Glennan and Illari 2017 suggest minimal definitions of mechanism that resemble each other. Andersen 2014 argues against a unified concept of contemporary mechanism.

[2] See Schiemann 1997 for the orientation to physical mechanics, Mumford 1981 for the machine metaphor, Dijksterhuis 1956 for the mathematization of scientific knowledge.

[3] Nicholson 2012, 154.

[4] For example, Glennan 1992, 12ff., Craver and Darden 2005, and Bechtel 2006, 20ff..

[5] See n. 1.

cal types. Before examining the ontological commonality more closely, I would like to mention as differences those aspects which are characteristic of the early modern, but not of contemporary, mechanism: the limitation of the *explanans* to matter in motion and the associated effort to reduce all phenomena to this type of change. Contemporary mechanism, by contrast, recognizes both reductionist and non-reductionist explanations without being restricted to matter in motion.[6]

The ontological types of early modern mechanistic philosophy are a function of the different ways in which they relate the concepts of matter and force. Matter is conceived as the substance in which location-changing movement takes place. It can be divided either discretely or continuously into segments, but it must be inherently unchangeable and be differentiated at most in the purely quantitative attributes of its particular form— i.e. size and shape. Forces may be responsible for cohesion among the various material particles, for their gravity, and for the movements, or kinds of movements, they make. "Monistic" views either reject an independent concept of force (e.g., Robert Boyle and Huygens) or, conversely, explain all properties of matter as effects of forces (e.g., Gottfried Wilhelm Leibniz and Immanuel Kant). An intermediate position between these two extremes is taken by the "dualistic" mechanism that conceives of matter and force as irreducible basic concepts (e.g., Newton and Roger Boscovich).[7]

A similar classification can now be made for two of the three groups of the main present-day proponents of mechanism. It refers to the relationship between the concepts of entity and activity. Both terms are defined contextually, and they are not used in entirely uniform ways. The term entity is sometimes used synonymously with that of part, the term activity sometimes synonymously with that of interaction between parts.[8] MDC describe the ontology of their conception explicitly as a dualist one and classify Glennan (1996) under substantivalism.[9] I would like to show in the following that this self-characterization is correct and that Glennan's monism concerning entities can be interpreted as a version of substantivalism. Furthermore, it must also be demonstrated that the conceptual pairs "entity/activity" and "matter/force" are at least structurally related. For this purpose, I will refer to the context of physical phenomena.

I will begin with a discussion of examples of the ontological difference in a narrow early modern version of mechanism (Sect. 3.2), then analyze the two ontologically related conceptions of contemporary mechanism (Sect. 3.3), and, in conclusion, highlight some aspects of the comparison between the early modern and contemporary conceptions of mechanism presented (Sect. 3.4).

---

[6] Williamson 2011.

[7] Schiemann 2009, 15ff. and 33ff.

[8] For example, Illari and Williamson 2012, and Glennan 2016.

[9] Machamer et al. 2000 (hereafter cited as "MDC 2000"), 4f.

3.2 Early Modern Mechanism

3.2.1 Monism in Christiaan Huygens

Under “materialistic” monism I understand the version of mechanism that rejects non-material forces as explanatory entities and recognizes pure contact between material bodies as the only possible form of natural interaction. The phenomena are explained in terms of pressure and impact processes between elementary bodies.[10] The first historically influential articulation of this mechanism was Robert Boyle’s conception of nature, though this can be described as mechanistic only in a broad sense. In the work of Christiaan Huygens, the second prominent representative of materialist monism, physical mechanics becomes the structuralizing principle.

In his conception of nature, Huygens builds directly on Boyle and he places the explanation of the phenomena of light and the weight of bodies at the center of his mechanism.[11] Specifically for the purpose of deducing gravity from mechanical centripetal forces, he postulates a fine-grained and weightless ethereal substance that rotates in a spherical vortex motion around the earth, pushing bodies toward the center of the earth.[12] According to Huygens, the space between the ether particles is empty, so that he can attribute the free mobility required for the calculations of mechanics to these particles.[13]

Huygens’ atomism ascribes hardness as a property to both the ether particles and the non-ethereal parts of which all bodies are composed.[14] Because matter is supposed to be the sole first cause of all phenomena, so that it cannot have different degrees of hardness, he has to posit the resistance caused by hardness as absolute.[15] Huygens ignores the dictate of intuitive representations that dominates early modern scientific discourse and postulates that elementary collisions are elastic despite the absolute hardness of the collision partners:

> Whatever may be the cause of hard bodies rebounding from mutual contact when they collide with one another, let us suppose that when two bodies, equal to each other and having equal speed, directly collide with one another, each rebounds with the same speed which it had before the collision (Huygens 1977, 574).

On this assumption, he provides a correct formulation of the laws governing elastic collisions, but not a correct explanation of gravity.[16]

Huygens’ success in deriving the laws of collisions comes at the cost of the non-intuitiveness of his concept of matter and contrasts with his failure to provide an

[10] See the definitions of Kirchner 1833, 212f., and Brugger (Ed.) 1950, 213.

[11] On Huygens’ mechanistic philosophy of nature, see Lasswitz 1890, Vol. 2, 341ff., Westman 1980, Snelders 1980 and Gabbey 1980. On the following, see Schiemann 1997, 95ff.

[12] Huygens 1896, 5ff.

[13] Loc. cit., 31f.

[14] Ibid. See Lasswitz 1890, Vol. 2, 360ff.

[15] Letter from Huygens to Leibniz of 11.7.1692, in Leibniz 1849ff., Vol. 2, 139.

[16] See Dühring 1873, 165ff. and Dugas 1957, 176ff.

explanation of gravity (a problem that remains unsolved to the present day). In his discussion of early modern corpuscularism, under which he classifies Huygens, Glennan interprets this failure as an early indication that gravity cannot be explained in mechanistic terms in principle.[17] In the case of fundamental forces such as gravity, according to Glennan, the causal understanding of the world as such reaches its limits with (not only early modern) mechanistic explanations formulated in causal terms.[18] I will return to the importance of the existence of a fundamental level for Glennan's and MDC's conceptions of mechanism below.

At this point, it should be noted that Glennan (in contrast to MDC) understands his conception as the result of a critical confrontation with the early modern variants of mechanism.[19] Huygens' mechanism also shows, according to Glennan, that the explanation of the phenomena must not be confined to mechanical pressure and impact processes.[20] He regards the liberation from this strict requirement as the decisive precondition for the triumph of Newton's mechanistic explanation of the movements of bodies in space.

3.2.2 *Isaac Newton's Dualism*

Fundamental to the early modern understanding of mechanical forces is the distinction between the uniform inertial motion of material bodies free of forces, on the one hand, and changes in this motion caused exclusively by forces, on the other.[21] If we assume that matter is initially in motion, then any subsequent change in motion of material bodies requires a measurable mechanical force, which, in dualistic and narrowly conceived mechanism, has the status of the only permissible cause for changes in nature—apart from the effects that can be attributed to the shapes assumed by atoms (adhering, interlocking, etc.). Forces can exercise effects without bodies having to touch each other because the forces, which in narrow dualist mechanism only act between material bodies, operate even when the material bodies in question are spatially separated. The relevant conception of matter is atomistic and the vacuum or the ether is where forces operate.[22]

---

[17] Glennan 1992, 13ff. and 138.

[18] Loc. cit., 174f.

[19] Andersen 2014, 281, points to Glennan's greater indebtedness to the history of mechanism in comparison to MDC (and Bechtel).

[20] Glennan 1992, 19ff.

[21] The assumption of inertial motion presupposes the existence of matter independent of forces. However, forces only work between material bodies. In this respect, dualistic mechanism exhibits an affinity with Cartesian metaphysics, in which non-human creatures do not have minds, but the mind occurs in the experienceable world only together with the human body. See n. 50.

[22] On this and the following, see, Thackray 1970, Westfall 1971, Freudenthal 1982 and Schiemann 2009, 35ff.

Newton's narrow mechanism is structured along the lines of his theory of gravitation. Gravity refers to the attractive force postulated by Newton that operates in empty space along the straight line connecting the centers of gravity of two macroscopic bodies with a value proportional to their masses and inversely proportional to the square of the distance between them. Because, according to Newton, nature "will [always] be very conformable to herself and very simple" (Newton 1704, 258) similar forces must also operate between small submicroscopic parts, of which he assumes all material things are composed.[23]

The essential attributes of the smallest parts are exactly the same as those of all things "that we handle" and that can be examined by simple experiments (Newton 1999, 795). For Newton, these are: "extension, hardness, impenetrability, mobility" and, finally, the "force of inertia" (ibid.). In contrast to gravity, whose strength depends on distance, inertia is an invariant attribute.[24] It is a "passive principle" through which "there never could have been any motion in the world" (Newton 1704, 258). For Newton, an "other", "active principle" is realized in the effects of forces (ibid) that, together with the passive principle, collectively constitute a dualism upon which all natural phenomena rest.

Newton was famously reticent when it came to statements about the nature of gravity. He thought, however, that an explanation that goes beyond his mechanical theory was desirable. He himself considered both non-mechanical origins[25] and mechanistic-materialistic causes.[26] Without having found a solution, he made the decisive determination for dualistic mechanism at least in a negative sense. In the "Rules for the Study of Natural Philosophy" that introduce the third book of the *Principia* he writes: "Yet I am by no means affirming that gravity is essential to bodies" (Newton 1999, 796). In this way, matter and force are distinguished from each other as two principles of the mechanistic explanation of nature in a manner characteristic of this tradition. They are juxtaposed as a passive and an active principle and mutually condition each other.[27]

Glennan argues that the ontological difference in Newton's mechanism represents methodological progress. Newton, he argues, recognized that a mechanistic explanation of bodily motion was possible even though the nature of gravity itself remained unexplained.[28] Following Newton, Glennan advocates a hierarchically structured theory of levels: In this case, the lower level is formed by fundamental processes that induce gravitation and that Huygens wanted to understand as a presupposition of natural explanation; the upper level consists of the perceptible and measurable mechanical effects of gravity and movements of bodies. According to

[23] Loc. cit., 261.

[24] Newton 1726, 388.

[25] Loc. cit., 237.

[26] Newton 1704, 385.

[27] On the polarity of the two principles in Newton, see Freudenthal 1982, 40ff. and 265ff.

[28] Glennan 1996, 20f. Similarly, Bechtel 2006, 20ff., describes the history of mechanism from the beginning of the modern era to the present day as a process of progressive detachment from initially restrictive mechanical requirements.

Glennan, we owe Newton the insight that explanations of the phenomena of the upper level are independent from those of phenomena of the lower level. Remarkably, however, Glennan, unlike Newton, rejects an independent concept for the forces or active principles that exert effects in the upper level, as I will explain in the next section.

3.3 Contemporary Mechanism

3.3.1 Monism in Stuart S. Glennan

Glennan's mechanism has been characterized in various places as monism. Thus Ivarola et al. (2013) describe Glennan's position as a "monist position according to which mechanisms are composed of entities interacting in a stable way" (Ivarola et al. 2013, 22).[29] Glennan has maintained the monistic position in essence since his first formulation of mechanism in 1992. Modifications he has made involve changes in the definition of the interactions between the entities to which interactions are attributed as properties.

Entities have an explanatory character. As part of a complex system, they give rise to its properties through their interactions.[30] From the beginning Glennan also uses the expression "part" for the concept of entity.[31] It is conceived in a broad sense in order to be able to satisfy as general a claim to validity as possible:

> Parts may be simple or complex in internal structure, they need not be spatially localizable, and they need not be describable in a purely physical vocabulary. … The parts of mechanisms must have a kind of robustness and reality apart from their place within that mechanism. Care must be taken so that parts are neither merely properties of the system as a whole nor artifacts of the descriptional vocabulary. I shall summarize these restrictions by saying that parts must be objects (Glennan 1996, 53).[32]

Entities encompass far more than just the material bodies that served as *explanans* in early modern mechanism. Depending on the context, they can render an independent concept of force superfluous or also designate objects that correspond to this concept. An independent concept of force is dispensable insofar as forces can be described through changes in the properties of entities. Glennan initially defined

---

[29] Correspondingly also Gebharter and Kaiser 2014, 63, Kaiser and Krickel 2016, 22, and Kaiser 2017, 116 and 121f. According to Torres 2009, 238, Glennan could also dispense with interactions as a matter of ontology: "Glennan's ontology posits entities as ontologically basic with interactions serving a solely descriptive purpose in mechanism models." However, this seems to contradict Glennan's own account of his position (see quotation to n. 31). An example of Glennan's own monistic definition of mechanism can be found in Glennan 2002, S352: "mechanisms are collections of parts."

[30] See the definitions of mechanism in Glennan 1992, 24, and Glennan 1996, 52.

[31] For example, Glennan 1992, 30: "This description would lead to a decomposition of the system in which the parts were electrons, molecular lattices, or other such entities."

[32] Cf. Glennan 1992, 31f.

changes in properties in terms of the concept of law. Taking the example of Newton's law of gravitation, he shows how force dissolves, as it were, in the law-governed changes in location involved in the movements of parts.[33] Within the framework of Glennan's approach, Newton's dualism in this way becomes a form of monism.[34] However, the broad understanding of the concept of entity can also refer directly to interactions, as Glennan explains using the example of the electromagnetic field, which describes phenomena that satisfy the early modern concept of force.[35]

Glennan subsequently replaced the concept of law with the more general concept of "invariant, change-relating generalizations" that he adopted from Jim Woodward's theory of causation.[36] By including the predicate "change-relating" in the definition of interactions, he reinforces their characterization as properties of entities on which change operates. At the same time, the new definition addresses an objection against using the concept of law also made by MDC (2000), according to which the regularities of mechanisms cannot be described in terms of laws in certain object domains, such as those of molecular biology and of neurobiology.[37] Invariant generalizations can, but need not, be laws. They do not claim the exceptionless validity of laws.[38]

Like the broad concept of entity, the new definition of interactions follows the quest for a validity claim that tries to capture all scientific objects in principle.[39] The universality of Glennan's mechanism also subserves the abovementioned hierarchically structured theory of levels, which takes up the idea, going back to ancient atomism and revived in early modern times, of explaining phenomena (in the upper level) in terms of the mechanisms underlying them (in the lower level). According to Glennan, every mechanism can be the object of a deeper-level explanation until a fundamental level has been reached for which there is no further explanation.[40] As examples of presumably fundamental interactions, Glennan cites physical forces such as gravity or electromagnetic interactions.[41] With early modern materialist monism, which I introduced with reference to Huygens, Glennan shares the reductionist quest for a fundamental explanation. Unlike Huygens, however, Glennan

[33] Glennan 1992, 38f.

[34] In this way, the concept of force is traced back to matter only in a formal sense, however, but is not explained in the sense of early modern materialist monism. Gustav Kirchhoff defended a similar approach in his mechanics (Kirchhoff 1876).

[35] Glennan 1992, 34ff. The forces described by the electromagnetic field were also an object of mechanism in the early modern period. Thus, Johannes Kepler assumed that magnetic forces operate between moving masses, as Newton was aware (Wilson 2002, 204f). Pierre Gassendi tried to explain these forces in mechanistic terms (Fischer 2013).

[36] Glennan 2002, S344ff., Woodward 2000.

[37] MDC 2000, 7.

[38] Glennan 2002, S345.

[39] Since 1992, Glennan defends a claim to validity for his theory of explanation that encompasses scholary knowledge as a whole.

[40] Glennan 1992, 138ff., Glennan 2002, 18. See Torres 2009, 238 and Williamson 2011, 429ff. However, Glennan 2016, 814, concedes that there also cannot be any fundamental level.

[41] Glennan 1992, 138. Cf. n. 17 and Glennan 2002, 18.

affirms the autonomy of the non-fundamental explanations for which Newton's dualistic mechanism provides an example. Glennan can integrate the pair of concepts that occur in dualism into his broad concept of entity.

In recent publications, Glennan understands interactions as a subset of activities, whereby it remains unclear whether he adopts MDC's concept of activity and thus makes a concession to their dualistic mechanism, or whether, on the contrary, he thinks that he can integrate elements of their approach into his monism.[42] On the one hand, this openness may be due to the fact that ontological differences increasingly play only a subordinate role for the application of the models of mechanistic explanation to scientific practice.[43] On the other hand, dualism claims with some justification that it does better justice to the phenomena for practical purposes than monism, as we will see.

3.3.2 Dualism in Peter Machamer, Lindley Darden, and Carl F. Craver (MDC)

MDC (2000) justify their dualism as an attempt to overcome the one-sidedness of monistic approaches, which they identify as substantivalism and process ontology.[44] By subsuming Glennan (1996) under substantivalism, they relate the latter to contemporary mechanism. Substantivalism "confine[s] [its] attention to entities and properties, believing that it is possible to reduce talk of activities to talk of properties and their transitions" (MDC 2000, 4).[45] This characterization is incorrect insofar as Glennan's context-relative concept of entity does not assume the immutability of substances. Nevertheless, it has a certain justification insofar as the concept (following an early modern tradition) also includes a persistent fundamental or substantial level of objects that resists explanation.

Process ontologists "reify activities and attempt to reduce entities to processes" (loc. cit., 5). The first part of this characterization ("reify activities") describes a concept that resembles the early modern monism of force, provided that forces fall under the concept of activity. By demarcating their dualism from two monistic conceptions, MDC cover, at least in a rudimentary way, a spectrum that exhibits a striking resemblance to the three ontological types of early modern mechanism

[42] Glennan 2016, 799, referring to Glennan (forthcoming).

[43] Glennan 2016, 799, does not attach any special importance to the difference between his concept of interaction and the concept of activity that he still criticized in Glennan 2010, 320ff. Having distanced himself in MDC 2000 from Glennan's and Bechtel's concept of interaction, Craver — in Craver and Tabery 2015, stressing the importance of scientific practice for mechanism — declares the differences between the three groups of the main representatives to be bridgeable.

[44] MDC's Mechanism has been characterized in various places as dualism, e.g. Tabery 2004, 2, Torres 2009, 233ff., Illari and Williamson 2013, 69ff., and Kaiser 2017, 116 and 121–124.

[45] According to Glennan 2010, 320f., MDC not only classify his own conception under substantivalism but also that of Bechtel and Richardson 1993.

(see section 1). However, process ontology, of which Rescher's process metaphysics is cited as an example,[46] occupies only a relatively marginal position in contemporary discussions of mechanism.[47]

MDC object that substantivalism does not take sufficient account of the productivity captured by the concept of activity with which changes in properties of entities are effected. [48] Entities do not bring about the changes in their own properties. What in Glennan has the character of a black box[49] is covered by the concept of activity in MDC. They criticize process ontology in more specific terms than substantivalism. Since there are no activities unrelated to entities in the field of neurobiology and molecular biology on which MDC's mechanism concentrates, process ontology, they argue, is not applicable because it denies the necessity of this relation. MDC's concept mechanism assumes that

> Mechanisms are composed of both *entities* (with their properties) and *activities*. Activities are the producers of change. Entities are the things that engage in activities. ... Entities and activities are correlatives. They are interdependent. ... There are no activities without entities, and entities do not do anything without activities (MDC 2000, 3, 6 and 8).

They juxtapose a passive and an active principle in a way comparable to Newton's early modern dualism. However, the assertion that entity and activity are indissolubly linked ("interdependent") is at odds with the dualist idea that the two principles distinguished are also independent of each other.[50] By denying that activities occur without entities and that entities occur without activities, dualism moves closer to monism.[51]

Notwithstanding the interdependence of entity and activity, the associated concepts remain clearly distinct from each other. The concept of entity is at first circumscribed in a similarly vague way to Glennan's conception as "parts in the mechanism with their various properties" (Craver and Darden 2013, 16), but is then restricted primarily through the definition of the concept of activity. This includes not only objects of explanation of different disciplines but also their historical transformations. To the concept of activity belong, for example, geometric-mechanical activities, which describe both the interactions between the corpuscles of early modern materialistic mechanism and the "fitting of a neurotransmitter and a post-synaptic

[46] Rescher 1996.

[47] See Williamson 2011, Illari and Williamson 2013 and Andersen 2014. Levy and Bechtel 2016, 14, nevertheless identify the orientation to process ontology as an option for the future development of mechanism ("mechanism 2.0").

[48] MDC 2000, 5.

[49] Tabery 2004, 10f.

[50] In modern dualism, this independence is not entirely symmetrical. See n. 21. Descartes' dualism postulates bodies without mind (non-human organisms) and mind without a body, i.e. human souls, but in human beings minds do not occur without bodies. For contemporay mechanism, Illari and Williamson 2012, 130f., show that activities are conceivable without entities and entities without activities.

[51] Glennan 2010, 321 points out this proximity when he treats the concepts of interaction and activity as interchangeable. "Where MDC speak of entities and activities, ... Glennan speak[s] of parts and interactions" (ibid.).

receptor" (MDC 2000, 14). But gravity and other physical interactions (e.g., electrostatic attraction and repulsion, magnetism) are also examples of activities[52] that Glennan ascribes at the same time to the entities or their interactions.

Like Glennan, MDC also assume that mechanistic explanation exhibits a hierarchical level structure.[53] Unlike him, however, they do not assume that there is a fundamental level underlying all mechanistic explanations. The range of mechanistic explanation is determined in pragmatic terms and comes to an end where the next-lower level is irrelevant to the epistemological interest.[54] MDC radicalize Newton's insight of the independence of higher-level explanations.

In addition to its conceptual innovation over other approaches (specifically, substantivalism and process ontology), MDC cite "descriptive adequacy" as a further justification for dualism (loc. cit., 8ff.).[55] In this way they emphasize the special importance that their approach attaches to the relation to scientific practice. Accordingly, they demonstrate the applicability of dualism using textbook examples of neurobiologists.[56]

3.4 Concluding Comparative Remarks

The early modern pair of concepts "matter" and "force" is structurally related as regards the contrast it draws to the contemporary conceptual pair "entity" and "activity." There are also overlaps at the level of content, in that the concept of matter is contained in that of entity and the concept of force is subsumed in part into the concept of (monistic) entity, in part into the (dualistic) concept of activity.

Through their reference to the context of mechanics, the early modern basic concepts were conceived much more narrowly than the basic concepts of contemporary mechanism. Contemporary mechanism has come closer to fulfilling the shared quest of the historically widely separated conceptions to achieve a uniform method of natural scientific knowledge simply because its basic concepts are wider in scope. However, this is also one of the reasons why the contrast between monism and dualism has become weaker. Unlike the materialistic monism of the early modern era, contemporary monism is no longer forced to explain interactions between the entities. There was a clearer separation between the concept of matter of the early modern materialistic monists and that of force than in contemporary monism, as far as this was dealt with here. The reduction task facing the early modern monists was correspondingly demanding—or rather, unrealizable. Glennan, by contrast, can integrate interactions into his broad concept of entity.

[52] Craver 2007, 64.

[53] MDC 2000, 4, 7, and 13f.

[54] Loc. cit., 13.

[55] As motivation for the dualistic approach to activities they specify ontological, descriptive, and epistemic adequacy (loc. cit., 4)—the last of which I have not discussed.

[56] According to Andersen 2014, 275, MDC belong to the group of "mechanisms as integral to scientific practice," Glennan, by contrast, to the group of "mechanisms as an ontology of the world."

Ontological differences played a much greater role in the early modern discourse about science than in contemporary philosophy of science.[57] This has also contributed to leveling the difference between contemporary monistic and dualistic approaches. All the more remarkable is the persistence of ontological questions, albeit in an attenuated form.

Present-day monism is associated with a form of reductionism that—similar to early modern monism—aims at a fundamental level. The characterization of this level makes use of a non-intuitive terminology. In Huygens it took the form of the mathematical and physical laws of the idealized elastic collision; Glennan accepts that the causal understanding of the world, which is otherwise indispensable, may fail at the fundamental level, assuming it exists.

Contemporary mechanism as a whole—as Glennan correctly emphasizes—is indebted to early modern dualism for the insight that successful explanations are possible even if the underlying processes are not yet understood.[58] As in the early modern period, contemporary dualistic mechanism enjoys the advantage over its monistic counterpart that its terminology is more closely related to scientific practice. The phenomena of the world seem to have a widespread dualistic character.[59] Dualism can provide a basis for good explanations even without having been explained itself. But monism, on the other hand, has the advantage that it remains an option for explaining dualism.

Acknowledgement I want to thank Stuart Glennan for his comments to an earlier version and Ciaran Cronin for his translation of the German version of this paper.

References

Andersen, H. 2014. A Field Guide to Mechanisms: Part I and II. *Philosophy Compass* 9: 274–293.

Bechtel, W. 2006. *Discovering Cell Mechanisms: The Creation of Modern Cell Biology*. Cambridge: Cambridge University Press.

Bechtel, W., and R.C. Richardson. 1993. *Discovering Complexity: Decomposition and Localization as Strategies in Scientific Research*. Princeton: Princeton University Press.

Brugger, W., ed. 1950. *Philosophisches Wörterbuch*. Freiburg: Herder.

---

[57] According to Cassirer, what sets early modern mechanism apart is the "logical primacy" of the traditional conception of substance (Cassirer 1923, 8). Today, by contrast, the ontological perspective is just one aspect among others, see n. 42 and 54 and Glennan 2016, 796. At the same time there is an ontological continuity of mechanistic explanations from early modern science to current scientific practice, see Falkenburg in Chapter I.4 of this volume.

[58] This can be interpreted as a renunciation of the ontological completeness of explanations. From an epistemic perspective, mechanism during the nineteenth century in addition relinquished the claim to absolute truth when mechanistic explanations are posited as having only hypothetical validity. See Schiemann 2009.

[59] This is also suggested by the physical theories of the very small, which (in the standard model) divide all elementary particles into the two classes of material particles (Fermions) and interaction particles (Bosons). For an introduction, see Carroll 2013 and Hauschild 2016.

Carroll, S. 2013. *The Particle at the End of the Universe: How the Hunt for the Higgs Boson Leads Us to the Edge of a New World*. London: Oneworld.
Cassirer, E. 1923. *Substance and Function. Chicago*. London: Open Court.
Craver, C.F. 2007. *Explaining the Brain: Mechanisms and the Mosaic Unity of Neuroscience*. Oxford: Clarendon Press.
Craver, C.F., and L. Darden. 2005. Introduction: Mechanisms in Biology. *Studies in History and Philosophy of Biological and Biomedical Sciences* 36C (2): 233–244.
———. 2013. *In Search of Mechanisms: Discovery Across the Life Sciences*. Chicago: University of Chicago Press.
Craver, C.F., and J.G. Tabery. 2015. Mechanisms in Science. In *Stanford Encyclopedia of Philosophy*, ed. E.N. Zalta. https://plato.stanford.edu/entries/science-mechanisms [07.04.2017].
Dijksterhuis, E.J. 1956. *Die Mechanisierung des Weltbildes*. Trans. H. Habicht. Berlin: Springer.
Dugas, R. 1957. *A History of Mechanics*. London: Routledge & Paul.
Dühring, E.K. 1873. *Kritische Geschichte der allgemeinen Principien der Mechanik*. Berlin: T. Grieben.
Fischer, S. 2013. Pierre Gassendi. In *Stanford Encyclopedia of Philosophy*, ed. E.N. Zalta. https://plato.stanford.edu/entries/gassendi [07.04.2017].
Freudenthal, G. 1982. *Atom und Individuum im Zeitalter Newtons*. Frankfurt am Main: Suhrkamp.
Gabbey, A. 1980. Huygens and Mechanics. In *Studies on Christiaan Huygens: Invited Papers from the Symposium on the Life and Work of Christiaan Huygens*, ed. Bos et al., 166–169. Amsterdam, 22–25 August 1979, Lisse: Swets & Zeitlinger.
Gebharter, A., and M.I. Kaiser. 2014. Causal Graphs and Biological Mechanisms. In *Explanation in the Special Sciences: The Case of Biology and History*, ed. M.I. Kaiser et al., 55–86. Dordrecht: Springer.
Glennan, S.S. 1992. *Mechanisms: Models, and Causation*. Ph.D. dissertation. Chicago: University of Chicago.
———. 1996. Mechanisms and the Nature of Causation. *Erkenntnis* 44 (1): 49–71.
———. 2002. Rethinking Mechanistic Explanation. *Philosophy of Science* 69 (S3): S342–S353.
———. 2010. Mechanisms. In *The Oxford Handbook of Causation*, ed. H. Beebee, 315–325. Oxford: Oxford University Press.
———. 2016. Mechanisms and Mechanical Philosophy. In *The Oxford Handbook of Philosophy of Science*, ed. P. Humphreys et al., 796–816. Oxford: Oxford University Press.
———. forthcoming. *The New Mechanical Philosophy*. Oxford: Oxford University Press.
Glennan, S.S., and P. Illari. 2017. Varieties of Mechanisms. In *The Routledge Handbook of Mechanisms and Mechanical Philosophy*, ed. S. Glennan and P. Illari, 91–103. New York: Routledge.
Hauschild, M. 2016. *Neustart des LHC: CERN und die Beschleuniger: Die Weltmaschine anschaulich erklärt*. Wiesbaden: Springer Fachmedien.
Huygens, Ch. 1896. *Abhandlung über die Ursache der Schwere*. Trans. and ed. R. Mewes. Berlin: Albert Friedländer's Druckerei.
———. 1977. *The Motion of Colliding Bodies*. Trans. R.J. Blackwell. In *Isis* 68(4): 574–597.
Illari, P., and J. Williamson. 2012. What Is a Mechanism? Thinking about Mechanisms 'across' the Sciences. *European Journal for Philosophy of Science* 2 (1): 119–135.
———. 2013. In Defence of Activities. *Journal for General Philosophy of Science* 44 (1): 69–83.
Ivarola, L., et al. 2013. Expectations-based Processes – An Interventionist Account of Economic Practice: Putting the Direct Practice of Economics on the Agenda of Philosophy Economics. *Economic Thought* 2 (2): 20–32.
Kaiser, M.I. 2017. The Components and Boundaries of Mechanisms. In *The Routledge Handbook of Mechanisms and Mechanical Philosophy*, ed. S. Glennan and P. Illari, 116–130. New York: Routledge.
Kaiser, M.I., and B. Krickel. 2016. The Metaphysics of Constitutive Mechanistic Phenomena. *The British Journal for the Philosophy of Science* 68 (3): 1–35.

Kirchhoff, G. 1876. *Vorlesungen über Mechanik. Bd. 1 der Vorlesungen über mathematische Physik*, ed. W. Wien. Leipzig: B.G. Taubner, fourth edition, 1897.
Kirchner, F. 1833. *Wörterbuch der Philosophischen Grundbegriffe*. Heidelberg: Georg Weiss Verlag.
Lasswitz, K. 1890. *Geschichte der Atomistik vom Mittelalter bis Newton*. 2 vols, reprint, Darmstadt: Wissenschaftliche Buchgesellschaft, 1963.
Leibniz, G.W. 1849 ff. *Mathematische Schriften*, ed. C.J. Gerhardt. 7 vols. Berlin: H. W. Schmidt.
Levy, A., and W. Bechtel. 2016. Towards Mechanism 2.0. Expanding the Scope of Mechanistic Explanation. In *PSA 2016. The 25th Biennial Meeting of the Philosophy of Science Association*. http://philsci-archive.pitt.edu/12567/ [05.01.2018].
Machamer, P.K., L. Darden, and C.F. Craver. 2000. Thinking about Mechanisms. *Philosophy of Science* 67 (1): 1–25 (abbreviated as "MDC 2000").
Mumford, L. 1981. *Mythos der Maschine: Kultur, Technik und Macht*. Frankfurt am Main: Fischer Taschenbuch Verlag.
Newton, I. 1704. Opticks. In Newton, I. 1779 ff. *Opera omnia*, ed. S. Horsley, Vol. IV, 1–264. London: Joannes Nichols.
———. 1726. *Philosophiae Naturalis Principia Mathematica*. The third Ed. with variant readings. Ass. and ed. A. Koyré and I.B. Cohen. 2 vols. Cambridge: Cambridge University Press, 1972.
———. 1999 ff. *The Principia: Mathematical Principles of Natural Philosophy. A New Translation*. Trans. I.B. Cohen and A.Whitman. Berkeley: University of California Press.
Nicholson, D.J. 2012. The Concept of Mechanism in Biology. *Studies in History and Philosophy of Science, Part C* 43 (1): 152–163.
Rescher, N. 1996. *Process Metaphysics: An Introduction to Process Philosophy*. Albany: SUNY Press.
Schiemann, G. 1997. *Wahrheitsgewissheitsverlust: Hermann von Helmholtz' Mechanismus im Anbruch der Moderne. Eine Studie zum Übergang von klassischer zu moderner Naturphilosophie*. Darmstadt: Wissenschaftliche Buchgesellschaft.
———. 2009. *Hermann von Helmholtz's mechanism: The Loss of Certainty: A Study on the Transition from Classical to Modern Philosophy of Nature*. Dordrecht: Springer.
Snelders, H.A.M. 1980. Christiaan Huygens and the Concept of Matter. In *Studies on Christiaan Huygens: Invited papers from the Symposium on the Life and Work of Christiaan Huygens*, ed. Bos et al., 104–125, Amsterdam, 22–25 August 1979, Lisse: Swets & Zeitlinger.
Tabery, J. 2004. Synthesizing Activities and Interactions in the Concept of a Mechanism. *Philosophy of Science 71* (1): 1–15.
Thackray, A. 1970. *Atoms and Power: An Essay on Newtonian Matter-Theory and the Development of Chemistry*. Cambridge, MA/London: Harvard University Press.
Torres, P. 2009. A Modified Conception of Mechanisms. *Erkenntnis 71* (2): 233–251.
Westfall, R.S. 1971. *Force in Newton's Physics*. London/New York: Macdonald and Elsevier.
Westman, R.S. 1980. Huygens and the Problem of Cartesianism. In *Studies on Christiaan Huygens: Invited papers from the Symposium on the Life and Work of Christiaan Huygens*, ed. Bos et al., Amsterdam, 22–25 August 1979, Lisse: Swets & Zeitlinger.
Williamson, J. 2011. Mechanistic Theories of Causality Part I. *Philosophy Compass* 6 (6): 421–432.
Wilson, C. 2002. Newton and Celestial Mechanics. In *The Cambridge Companion to Newton*, ed. I.B. Cohen, 202–226. Cambridge: Cambridge University Press.
Woodward, J. 2000. Explanation and Invariance in the Special Sciences. *The British Journal for the Philosophy of Science* 51 (2): 197–254.

Chapter 4
Mechanisms, Explanation and Understanding in Physics

Dennis Dieks

Abstract The Scientific Revolution is often associated with a transition to a "mechanistic" world view. However, "mechanization" is not the term that best captures the distinctive nature of modern physics: "mathematization" would be a better characterization. Modern physics attempts to find mathematical relations between quantities, and does not require that these relations be interpreted in terms of mechanisms. Moreover, in modern physics there are cases in which it is unnatural to give the mathematical formalism a mechanistic interpretation, even if "mechanistic" is broadly construed. Both on the level of ontology and that of explanation physics turns out to be more general and liberal than what is suggested by the catchphrase that physics explains by identifying mechanisms. Although mechanistic explanation remains an important conceptual tool, in particular for achieving understanding, it is not the only one available and cannot lay claim to fundamentality.

4.1 Introduction

In his book "The Mechanization of the World Picture" (Dijksterhuis 1961), the historian of science E.J. Dijksterhuis famously described the transition from ancient and medieval to modern science, in particular physics, as the replacement of occult qualities by clear and empirically accessible "mechanical" concepts like the size, velocity and acceleration of particles. Newtonian mechanics, the culmination of the Scientific Revolution, established mathematically formulated laws between quantities of this sort. The term "mechanization" seems apt for this transition, and it is true that the treatment of physical problems on the basis of mechanics became an ideal

D. Dieks (✉)
History and Philosophy of Science, Utrecht University, Utrecht, Netherlands
e-mail: d.dieks@uu.nl

B. Falkenburg, G. Schiemann (eds.), *Mechanistic Explanations in Physics and Beyond*, European Studies in Philosophy of Science 11,
https://doi.org/10.1007/978-3-030-10707-9_4

of physical science in the seventeenth, eighteenth and nineteenth centuries. However, as Dijksterhuis himself already noted in the epilogue of his book (Dijksterhuis 1961, p. 498), one should bear in mind that

> the science called mechanics had emancipated itself in the 17th century from its origins in the study of machines, and had developed into an independent branch of mathematical physics, dealing with the motion of material objects and finding in the theory of machines only one of its numerous practical applications.

As Dijksterhuis makes clear in his epilogue, with hindsight the fundamental contrast between ancient and medieval physics on the one hand and (early) modern physics on the other is not that the former sometimes uses explanations involving teleology or analogies with organisms whereas the latter models processes with concepts that come from the world of machines. Rather, the basic difference is that while ancient and medieval physics occasionally used mathematical tools, modern physics is essentially mathematical, defining core concepts in a mathematical way and formulating laws in mathematical language. Mathematical reasoning is by its nature abstract, and it is not self-evident that the metaphor of a "machine" will always be natural or even applicable for mathematically described processes, not even for processes within the domain of the science of mechanics itself.

The novel mathematical frameworks that were invented for the treatment of mechanical problems in the eighteenth and nineteenth centuries strengthen this point. As we shall see, these new approaches have led in the direction of growing abstraction and have made thinking in terms of what we intuitively would call "mechanisms" less than obvious. The development of new branches of physics in the nineteenth and twentieth centuries have further contributed to the general picture of increasing abstraction and distance from everyday intuition, also on the level of explanation.

Still, even though mechanistic explanations are not always the most usual and natural, one could hold on to the idea that such explanations are *possible in principle*, and provide a kind of basic understanding of physical processes. As we shall discuss in Sect. 4.3, it is indeed true that mechanical models can very often be constructed in physics—this was proved by Poincaré at the end of the nineteenth century. However, this existence in principle depends on a rather trivial underdetermination argument, and it is far from clear that it carries epistemological weight.

The fundamental status of mechanistic explanations was dealt a further blow by the advent of quantum mechanics in the twentieth century. It is a general trait of mechanistic explanations that they analyze the behavior of physical systems in terms of these systems' material constituents and the interactions between them: mechanistic explanations are meant to be "decompositional" (Psillos 2011). However, quantum theory has unsettled even such very general mechanistic conceptions. The very notion that a composite system can be fully analyzed in terms of the properties of its constituent parts and the relations between them has become debatable: according to standard interpretational ideas quantum theory attributes holistic features to physical systems. Thus even the most basic ingredients of the notion of a "mechanism" become moot.

Evidently, this unstable basis of the applicability of the concept of a "mechanism" has consequences for the status of mechanistic explanation in physics. Although there certainly are many cases in which mechanistic intuitions are helpful and provide understanding,[1] the associated pictures and concepts cannot lay claim to fundamentality. There are fundamental physical processes in which the usefulness of mechanistic modelling is far from obvious, and in which other types of reasoning appear more fitting. On balance, mechanistic explanations cannot claim to possess a privileged status.

As we shall argue, this points in the direction of a pluralist conception of explanation and understanding in physics, according to which contextual and pragmatic factors are important in deciding which conceptual framework is the most appropriate. Mechanistic explanation is one of the tools present in the "conceptual toolbox", but depending on the specifics of the problem case other explanatory strategies may be preferable.

4.2 Mechanics and Mechanisms

The intuitive attractiveness and power of descriptions in terms of mechanisms, in the original and literal sense of material objects whose parts interact via pushes and pulls, is beyond dispute. We are so familiar with the operation of pulleys, drive shafts and effects of collisions, that an analysis of complicated processes in such terms provides a strong conceptual grip on what is happening. Cartesian physics lived up to exactly this ideal of mechanistic explanation, which accounts for a great deal of its contemporary popularity. However, in the Scientific Revolution Cartesian physics proved just a brief phase: its framework was insufficiently flexible for the formulation of laws of motion of the kind finally stated by Newton (in particular, as Newton argued, the law of inertia cannot adequately be stated in a Cartesian material plenum without an independent non-material space-time background that serves to define what straight lines and equal time spans are).

Classical mechanics, in the form given to it by Newton, accordingly does not conform to the original mechanistic ideals. On the one hand, in addition to the traditional mechanical concepts of size, velocity and acceleration, there is a non-material arena, formed by space and time, that influences the evolution of physical processes; on the other hand, in addition to the familiar interactions via contact between impenetrable bodies action-at-a-distance forces are introduced. It is well known how Newton's contemporaries for this reason accused Newton of reintroducing mysterious occult qualities and of spoiling the progress that had been made towards clarity through a mechanistic understanding of the world.

Nevertheless, due to its enormous predictive success the new Newtonian framework before too long became the dominant scientific paradigm. A new norm for

[1] We associate "understanding" with qualitative insight in the behavior of physical systems, in the sense of De Regt and Dieks (2005).

"mechanistic" became thus established: mechanistic explanation came to signify the decomposition of a material system into constituent particles, the specification of interactions via Newtonian forces between these particles, and the proof that these interactions (defined against the background of a pre-given space-time) were able to predict the observed behavior of the total system.

The forces in the original Newtonian scheme are simple "inverse square" *central* forces, i.e. forces falling off with the inverse square of the mutual distance between the interacting particles ($1/r^2$) and directed along the straight line connecting them. However, at the end of the eighteenth and the beginning of the nineteenth centuries it was found that this simple type of interaction could not account for what happens in phenomena involving magnetism and moving electrical charges, so that more complicated interaction formulas had to be written down. For example, in order to accommodate the interaction between moving electrical charges within an action-at-a-distance framework one needs forces that are not directed along the line connecting the particles and that depend not only on the particle positions but also on their velocities and accelerations. As a result, the mechanistic ideal had to be adapted once again. Instead of requiring that a process should be explained in terms of localized particles and Newtonian central forces between them, it now became sufficient to give an analysis in which the total system is decomposed into material parts with mathematically stated force laws acting between them—in addition, new properties which did not possess an immediately obvious mechanical interpretation, like electrical charge, had to be accepted.

Meanwhile, it should be recognized that the adoption of this "Newtonian" type of explanation, in terms of parts dynamically producing the whole, depends on a particular perspective on classical mechanics, which is not the only possible one. The mathematically formulated theory does not unavoidably lead to the Newtonian picture of a history unfolding in time, in which physical systems at one instant work together, via their interactions, to generate the immediate future. To start with, the idea of the "production" of new situations from the old ones as time passes does not sit well with the formalism of mechanics, or even with the structure of mathematical physics in general. This relates to the notorious problem of the "'flow of time": time occurs in mathematical physics as a parameter in the same way as the spatial coordinates, which makes it impossible to define a preferred *now*. There is no privileged point on the time axis, just as there is no preferred *here*. A fortiori, there is no definable motion of a *flow* of time, just as there is no shifting *here* within the formalism.

Of course, the notions of "here" and "now" do become applicable once an external spatio-temporal viewpoint is introduced, for example connected to an observer *who makes use of the theory*. From the internal theoretical viewpoint both the concepts of *now* and *here*, and the notion of the flow of time, are merely indexical, deriving their meaning from a reference to such an external viewpoint, and not inherent in the theory itself. This may be taken as a first indication that the choice of explanations in terms of "productive mechanisms" itself has a pragmatic and contextual background, relating to the interests of the user of the theory.

In any case, interpretations of the theory of mechanics that do not start from the assumption that the theory describes how systems change while time flows are possible, and are moreover natural when we look at the formalism from an abstract point of view. Such interpretations view the universe as laid out not only in space, but also in time, as a four-dimensional "block"—the block universe, which comprises the whole of history without making a distinction between an absolute (as opposed to an indexically defined) Past, Present and Future.

Patterns of explanation that fit in with this "static" perspective were in fact already proposed in the eighteenth and nineteenth centuries. In these alternatives to the original Newtonian approach (associated with names like Maupertuis, Euler, Lagrange and Hamilton) one does not focus on instantaneous forces that change the present physical state, but rather asks which path will be followed by a mechanical system (e.g. a particle) if it is given that it finds itself at position x_1 at instant t_1 and is located at x_2 at another time t_2. The Principle of Least Action (or more generally the Principle of Stationary Action) states that among all possible continuous curves connecting x_1 and x_2 in the given time interval, the one actually realized minimizes (more generally: makes stationary) the "action" $S = \int_{t_1}^{t_2} Ldt$.

In this formula the function L is the *Lagrangian*, defined as the difference between kinetic and potential energy of the system: $L = T - V$. The kinetic energy T ($\frac{1}{2}mv^2$ for a point particle) will involve squares of velocities, the potential energy V will usually be only a function of positions, so that the total Lagrangian $T - V$ is a function of positions and velocities.

The Principle of Least Action may suggest that a system "chooses", from all logically possible evolutions between t_1 and t_2, the one that makes the action $\int_{t_1}^{t_2} Ldt$ minimal. This invites teleological patterns of explanation: the system, "knowing" that it will have to arrive at x_2 at time t_2, fulfils this task in the most economical way at its disposal.

Obviously, such anthropomorphic terminology, although not uncommon in the practice of physics, should not be taken seriously. Mathematically speaking, the Newtonian and Lagrangian approaches are equivalent: one can be derived from the other, so that arguments on the basis of the principle of least action need not introduce irreducibly novel ontological ingredients. Still, in explanations starting from the minimization of the action the focus is different than in Newtonian explanations: one now looks at the total path (if the system comprises more than one particle this path is defined in phase space), stretched out in time, and compares different possibilities. By contrast, in the Newtonian approach one focuses on the instantaneous state and computes how this state evolves in response to causal influences. This difference illustrates how the same mathematical theory (in this case classical mechanics) may be cast in various forms and how different patterns of explanation can become plausible depending on these different forms. In fact, the Lagrangian formulation is not the only non-Newtonian form that can be given to classical mechanics: the Hamiltonian and Hamilton-Jacobi formalism are still other alternatives, and there are more.

These more modern approaches in mechanics usually do not work with causal terminology (forces producing changes) and rely more on mathematical properties of the formalism. This opens up the possibility of new types of explanation, for example those based on the existence of *symmetries*. By way of illustration, if the action S (as defined above) does not explicitly depend on time (i.e. if time does not occur in the formula for S in addition to its implicit dependence on time via the coordinates and velocities—this expresses "symmetry under time translation"), it can be shown that the *energy* of the system remains constant over time (conservation of energy); if the action does not depend on position (symmetry under space translation) it follows that *momentum* is conserved. These examples illustrate Noether's theorem, which in a general and systematic way links symmetries to the conservation of physical quantities.

Summing up, what a mechanism is, and what a mechanistic explanation amounts to, is not completely set in stone in classical mechanics. In the course of history a development into the direction of more complicated and intuitively less immediately attractive "mechanisms" has proven necessary. A constant theme in this development (until the advent of quantum mechanics, about which more in a moment) has been the notion that a system should be decomposed in its constituent particles and that the whole should be understood on the basis of the dynamics of these parts. All these various mechanistic explanations make use of intuitively plausible causal terminology (forces, production, unfolding in time) and often provide understanding. However, within the same science of classical mechanics more abstract explanations (on the basis of variational principles, symmetries, or abstract properties of the mathematical structure) are possible as well, and actually occur more frequently in advanced treatments of the subject. So explanation by "mechanisms" is not inextricably bound up with the science of mechanics: mechanics is more flexible than that, and more neutral with respect to possible patterns of explanation.

4.3 Maxwell's Theory and Poincaré's Theorem

The eighteenth and nineteenth centuries saw the introduction of several new types of matter, "fluids", for the purpose of explaining phenomena in a number of relatively new disciplines: chemistry, the theory of heat, and most famous and important for our topic, electrodynamics. It was mentioned in the previous section that a Newtonian treatment of moving electrical charges meets with difficulties and requires the introduction of action-at-a-distance forces of a new and unusual sort. The introduction of the electromagnetic ether and the development of the field concept, culminating in Maxwell's theory of electrodynamics of 1865, cast new light on this subject. Instead of thinking in terms of action at a distance between particles, Maxwell proposed to conceptualize the interaction between charges as mediated by undulations in an underlying medium—waves that propagated, with a finite velocity, between the charges. This new "field-theoretic" framework was a huge success: Maxwell was able to unify electricity, magnetism and optics within the same theory. As he wrote (Maxwell 1865): "[It seems] that light and magnetism are affections of the same substance, and that light is an electromagnetic disturbance propagated through the field according to electromagnetic laws." Maxwell's new, general and comprehensive theory of electromagnetism made the earlier direct-action attempts obsolete.

In his 1873 *Treatise on Electricity and Magnetism* (Maxwell 1873) Maxwell presented his definitive treatment of the theory, in which electrodynamic quantities were represented by vectors, as still usual today: for example, $\vec{\mathbf{E}}$ stands for the electric field and $\vec{\mathbf{B}}$ for the "magnetic induction". In the rapidly following further development of the discipline these vectors came to stand for locally (i.e. per spatial point) defined forces, existing throughout the medium (the ether), so that we have a continuous "field" of electric and magnetic forces $\vec{\mathbf{E}}(x)$ and $\vec{\mathbf{B}}(x)$, with x indicating position. These forces are "felt" by localized charged particles that find themselves at the position x. As derived by Lorentz (see Darrigol 2000), the force exerted on a particle with electrical charge e and velocity $\vec{v}$, at position x in the field, is given by $\vec{F} = e\vec{\mathbf{E}}(x) + e\,\vec{v} \wedge \vec{\mathbf{B}}$ (with $\wedge$ denoting the outer product between two vectors).

The core of the theory developed in the Treatise (Maxwell 1873) is formed by the "Maxwell equations", which govern the dynamics of the electric and magnetic fields. After a long chain of arguments, Maxwell's presentation culminates in the demonstration that this dynamics can be put in *Lagrangian form*.

The final result is a theory in which we have field quantities defined throughout space and interrelated by a set of equations (the Maxwell equations). These fields exert influences on charged particles, and charged particles in turn influence the fields; this is all described in a rigorous and abstract mathematical way. Significantly, Maxwell states that in the final analysis these field quantities express the *mechanical state* of the underlying substance, the ether. But it is not worked out how exactly this should be fleshed out: in what way does the state of motion of the material ether generate electric and magnetic fields, and how should we envisage the interaction between charged particles and the moving parts of the ether?

In 1890 Henri Poincaré published a book, *Électricité et Optique* (Poincaré 1890), meant to explain Maxwell's theory to the French-speaking world. Poincaré starts his Introduction with the remark that for a French reader the first acquaintance with Maxwell's text will probably lead to a feeling of embarrassment and even distrust—as Poincaré states, this feeling will only disappear after much effort, and some eminent minds even keep it for ever.[2] According to Poincaré there are several reasons for this reaction: the French reader wants logic, consistency and precision, preferably in the form of a deductive system with a minimum number of clearly stated axioms, and Maxwell's work does not possess this form. But there is also another reason. Behind the world of experience the French reader will wish to see another world, consisting of matter with purely geometric properties, with atoms

[2] La première fois qu'un lecteur français ouvre le livre de Maxwell, un sentiment de malaise et souvent même de défiance se mêle d'abord à son admiration. Ce n'est qu'après un commerce prolongé et aux prix de beaucoup d'efforts, que ce sentiment se dissipe. Quelques esprits éminents le conservent même toujours.

that are point particles subjected to mechanical laws. Only then will he have the feeling that he has penetrated to the secret of the Universe.[3]

Poincaré expresses doubt about the philosophical tenability of this latter geometrical/mechanical requirement; he thinks that it concedes too much to our intuitive urge for easily visualizable pictures. Anyway, at this point in his Introduction Poincaré warns his readers that desires for an elegant axiomatic set-up and a simple and obvious mechanistic explanation will not be satisfied by Maxwell. However, and this is according to Poincaré meant to be Maxwell's most important general message to the reader of the Treatise, Maxwell *does* show that a mechanistic explanation of electric and magnetic phenomena is possible *in principle*.[4] To show that this conclusion is indeed contained in Maxwell's work, Poincaré proves a little theorem, still in the Introduction to his book (Poincaré 1890, ix–xiv).

The proof hinges on the fact that Maxwell's theory can be given a Lagrangian formulation, as emphasized in the Treatise. Quite generally, as Poincaré is going to show, theories with a Lagrangian formulation admit a mechanical model in which masses that interact via forces derivable from a potential can be made responsible for what the theory predicts on the observable level. The idea of the proof is simple. Any physical theory should make contact with experience, and should therefore operate with physical quantities, $q_1, q_2, \ldots, q_n$, that are accessible to measurement. If the theory can be put in Lagrangian form, this means that there exists a potential energy function $V(q)$ of the quantities $q_1, q_2, \ldots, q_n$, and also a kinetic energy function $T(q,\dot{q})$ of these quantities and their time derivatives, so that the Lagrangian $L=T-V$ can be formed.

Now, if there is to be a mechanical model, it should be possible to find p values of masses, $m_1, m_2, \ldots, m_p$, and p positions $x_1, x_2, \ldots, x_p$, of p particles that together determine the measurable quantities $q_1, q_2, \ldots, q_n$ and in turn are functions of these measurable quantities: $x_i=x_i(q_1, q_2, \ldots, q_n)$.[5] The kinetic energy, when expressed in the particle quantities, should have the usual particle form: $T(q,\dot{q})=\sum_{i=1}^{p}\frac{1}{2}m_i\dot{x}_i^2$. Since we are interested in the existence *in principle* of a mechanical model, no limit is set to the value of p, so that one may assume as many particles as one likes.

With this freedom in the number of particles it is always possible to satisfy the equation for the kinetic energy, $T(q,\dot{q})=\sum_{i=1}^{p}\frac{1}{2}m_i\dot{x}_i^2$: the number of unknowns, p can be made much greater than the number n of known quantities. We therefore have a case of mathematical underdetermination, and there will be infinitely many possible choices for the masses and positions.

---

[3] Derrière la matière qu'atteignent nos sens et que l'expérience nous fait connaître, il voudra voir une autre matière, la seule véritable à ses yeux, qui n'aurait plus que des qualités purement géométriques et dont les atomes ne seront plus que des points mathématiques soumis aux seules lois de la Dynamique…C'est alors seulement qu'il sera pleinement satisfait et s'imaginera avoir pénetré le secret de l'Univers.

[4] *Maxwell ne donne pas une explication méchanique de l'électricité et du magnétisme; il se borne a démontrer que cette explication est possible* [italics in original].

[5] More precisely, each particle position has three components in three-dimensional space, so that there are four unknowns associated with each particle.

Given any such choice, we can rewrite the potential energy V as a function of the particle positions, and the Lagrangian equations of motion become: $m_i\ddot{x}_i = -\frac{dV}{dx_i}$. So we have arrived at a theoretical scheme in which p moving particles, interacting via forces $-\frac{dV}{dx_i}$, fully reproduce the empirical predictions of the theory with which we started.

In other words, under very general conditions (the existence of a Lagrangian scheme) it is possible to find *many* mechanical models that lead to the exact same predictions as the given physical theory. These models will contain a number of point masses, interacting through local forces that derive from a potential. As said, there is no lack of such models: due to the underdetermination signalled above, if there is one such model, there is an infinity of them.[6]

Seen from this perspective, mechanical explanation is always possible—but it is cheap and resembles a sleight-of-hand. It seems an empty addition to what we could already understand in terms of the quantities q alone. As Poincaré discusses further in *La Science et l'Hypothèse* (Poincaré 1905, 196–197; 251–259), one might initially think that requirements concerning the form of the forces will give the concept of mechanical explanation more bite; for example, one could impose that the forces should be central, or expressible as fixed connections in the manner of Hertz, or perhaps reducible to the effects of direct particle collisions. But given the freedom to choose the number of particles p as large as one wishes, this will not help: the problem will remain underdetermined and there will still be infinitely many solutions.

Poincaré concludes that the choice for a mechanistic explanation will necessarily involve non-empirical, personal and pragmatic factors. He suggests that future physicists will no longer be interested in thinking about such things and will leave this to metaphysicians; and that in the end the reader of Maxwell's Treatise will see the artificial elements in the [mechanical] theoretical schemes that he once admired.[7]

4.4 Mechanisms and Quantum Mechanics

A common theme in the various forms of mechanistic explanation that we have considered is that they are decompositional (Psillos 2011): the behavior of a composite system is explained by reference to its material parts and the interactions between these parts. This has become the motivation for the "New Mechanicism" in the philosophy of science. This New Mechanicism is meant to be an elaboration of

[6]As Poincaré (Poincaré 1890, xiv) puts it: "*Si donc un phénomène comporte une explication méchanique complète, il en comportera une infinité d'autres qui rendront également bien compte de toutes les particularités révélées par l'expérience.*" [italics in original]

[7]Un jour viendra peut-être où les physiciens se désintéresseront de ces questions, inaccessibles aux méthodes positives, et les abandonneront aux métaphysiciens. …Le lecteur…finit par comprendre ce qu'il y avait souvent d'un peu artificiel dans les ensembles théoriques qu'il admirait autrefois (Poincaré 1905, 258–259).

and improvement on Salmon's causal scheme of explanation, according to which good explanations are ontologically grounded in the objectively existing causal structure of the world (Salmon 1984). The new mechanicists share this "ontic" commitment, but work out the details of the causal structure in a way that is slightly different from Salmon's original proposals, namely in terms of *mechanisms*, defined as complex, composite systems whose efficacy in performing a certain function can be understood on the basis of the concerted action of its parts. Glennan gives the following definition (Glennan 2002):

> a mechanism for a behavior is a complex system that produces that behavior by the interaction of a number of parts, where the interactions between parts can be characterized by direct, invariant, change-relating generalizations.

This definition accords well with the nature of the mechanisms that passed review in our historical sketch of classical mechanics. In particular, the behavior to be explained is *produced* by the *interactions* between the *parts*; and as Glennan explains, these parts must be *objects* with a high degree of robustness or stability, which are generally spatially localized. The interactions bring about changes in the properties of one part as a consequence of changes in the properties of another (Glennan 2002, 344). As Glennan adds concerning these interactions (Glennan 2002, 352), "events occurring at some point in space and time are explained as the consequence of the operation of causal mechanisms operating in that region of space and time. Our global evidence suggests that—quantum mechanics aside—causality is everywhere local."

The picture is that explanation by mechanisms is ontologically privileged as it latches on to the objective structure of the world: the world consists of composite objects whose behavior is produced by local interactions between localized component systems. It is important for this new mechanicism, as it was for older forms of mechanicism, that the interaction between any two components should be the same as in the case in which these components are the only systems present: the interactions should not be "holistic", depending on the behaviour of the complex system that is to be explained. The mechanistic intuition is that the global system should be reducible to its parts. No wonder then that Glennan added the clause "quantum mechanics aside" in the just-quoted statement: quantum mechanics is notorious for the problems it engenders for practically all of the mentioned ingredients of mechanistic explanation: according to quantum mechanics, physical systems need not be localized, interactions possess non-local aspects, and perhaps most important of all, the properties of a composite system can generally not be reduced to the properties of its parts.

There is one underlying reason for all these problems. Von Neumann already pointed out, in his seminal 1932 book on the mathematical structure of quantum mechanics (von Neumann 1932), that *the* central novel feature of quantum theory is that states of physical systems are to be represented by *vectors* in a state space (a Hilbert space), with the property that the *superposition* (sum) of any two such vector states again represents a realizable state. Accordingly, the structure of the quantum state space is radically different from what we are used to in classical physics.

For example, if we have two vector quantum states denoted by $|x_1\rangle$ and $|x_2\rangle$, meant to refer to a system at position x_1 and x_2, respectively, the sum state $\frac{1}{\sqrt{2}}(|x_1\rangle + |x_2\rangle)$ is again a *bona-fide* state—but this time we have a state that does not correspond to one definite localization. The superposition principle, saying that any two states may be superposed to form a new state in which a physical system can find itself, is responsible for most non-classical features of quantum mechanics.

In particular, the superposition principle explains why the state of a composite quantum system generally cannot be reconstructed from the states of its component parts. Suppose that we have a system C that consists of the two partial systems A and B; and suppose that possible states of A and B are $\{|\alpha\rangle_i\}$ and $\{|\beta\rangle_i\}$ (these are vectors in the Hibert spaces associated with A and B, respectively). In this situation, a simple composite state of C is

$$|\Psi\rangle = |\alpha\rangle_k \otimes |\beta\rangle_k, \tag{4.1}$$

which can be interpreted in a classical way: The system C (represented by $|\Psi\rangle$, a vector in the Hilbert space associated with C) consists of two components, A and B, with states $|\alpha\rangle_k$ and $|\beta\rangle_k$, respectively, and the properties of C supervene on these of A and B. The crucial point is that the superposition principle tells us that a superposition of states of the form (4.1) is also possible, which leads to a state of the form:

$$|\Psi\rangle = \sum_i c_i |\alpha\rangle_i \otimes |\beta\rangle_i, \tag{4.2}$$

where $|\alpha\rangle_i$ and $|\beta\rangle_i$ are again state vectors in the Hilbert spaces of A and B, respectively, and the coefficients c_i are complex numbers. In the situation represented by Eq. 4.2 the global system C is in a so-called pure state (represented by a vector in Hilbert space), but the partial systems A and B, taken by themselves, are in "mixed states". This can be intuitively understood from Eq. 4.2 because both A and B are associated with a whole range of state vectors ($\{|\alpha\rangle_i\}$ and $\{|\beta\rangle_i\}$, respectively) so that it is not implausible that they are best represented by a mixture of these states. Now, as von Neumann showed, it is a mathematical fact that $|\Psi\rangle$ fully determines the mixed states of A and B, but that the reverse is not true: The mixed states of the component systems, in a situation of the type represented by Eq. 4.2, do not fix the state of C, i.e. $|\Psi\rangle$. It follows from this that knowledge of all physical properties of A and B individually, and all possible outcomes of measurements performed on A and B by themselves, does not suffice to determine the state of C. There thus exists a certain holism in quantum mechanics: properties of a whole do generally not supervene on properties of the parts.

The answer to the question of whether one can think of composite quantum systems as being built up from parts that interact via local interactions (a "local model") relates to the just-sketched holism. As famously proved by Bell (1964), it is impossible in certain total states of the form (4.2) to reproduce, with a local model, the quantum mechanical predictions for the correlations between outcomes of measurements performed on A and B separately. The relevant quantum mechanical predictions have been impressively confirmed in many experiments, so that the conclusion is justified that nature is not correctly described by models with local interactions between parts—which clearly are *mechanisms* of the sort discussed earlier.

We already mentioned the general lack of localizability of quantum systems, which leads to another discrepancy between the quantum ontology and the ontology of local mechanisms. Since quantum states may be superpositions of states that are (more or less) localized, the resulting states can have very extended spatial domains. This is important for practical applications, as illustrated by the famous "double-slit" example, in which a single electron goes to two slits at the same time, in spite of a substantial distance between the slits. In the beginning days of quantum theory examples like this were mere thought experiments, but now they are routinely realized in laboratories and prove to be important for practical applications.[8]

The quantum world is therefore strange and radically non-classical, as emphasized in many accounts of the theory. Nevertheless, it is clear that if the theory is to be empirically adequate, classical patterns of behavior have to emerge in some situations—after all, there must be a reason that classical mechanics was successful for so long a time. If there were no classical limiting situations, classical physics would never have developed. The details of the classical limit of quantum mechanics are to some extent controversial, because they relate to interpretational issues (in particular, the measurement problem). However, there is a growing consensus that the process of "decoherence" is of vital importance here.

Decoherence occurs when a quantum system couples to its environment—usually an environment with very many degrees of freedom, which makes the process practically irreversible. The interaction with this environment is governed by the usual quantum mechanical evolution (the Schrödinger equation or a relativistic generalization of it); it is a case of ordinary quantum mechanical interaction. As we shall see in a moment, one of the effects of decoherence is that the effects of entanglement and superposition become hard to detect.

An entangled state has the general form shown in Eq. 4.2. Now suppose that there is an environment E that interacts with the system in the state $|\Psi\rangle$ of Eq. 4.2, such that E responds differently to the different terms in (4.2). This can be represented mathematically by the following evolution:

$$|\Psi\rangle |E_0\rangle = \sum_i c_i |\alpha\rangle_i \otimes |\beta\rangle_i |E_0\rangle \mapsto \sum_i c_i |\alpha\rangle_i \otimes |\beta\rangle_i |E_i\rangle. \tag{4.3}$$

[8] We here follow standard interpretational ideas, staying close to the standard Hilbert space formalism. The interpretation of quantum mechanics is notoriously controversial, and there are proposals that differ from the standard account. The difficulties mentioned in the text assume different forms depending on the interpretation that is being considered, but in any interpretation there remain holistic and non-local features. For example, in the Bohm version of quantum mechanics (Bohm 1952) particles *are* localized, but they interact via non-local forces of a holistic sort: the form of these action-at-a-distance forces depends on the quantum state of the composite object. So also here there is no supervenience of the whole on the parts.

In this formula the symbol $\mapsto$ represents the evolution: this evolution maps the initial state on the left hand side of the symbol into the final state on the right hand side. $|E_0\rangle$ is the initial environment state; the states $|E_i\rangle$ are the environment states that couple to the states $|\alpha\rangle_i \otimes |\beta\rangle_i$ of the composite object.

The crucial fact that makes decoherence so important is the following. When one performs a measurement on the composite system alone, after its interaction with the environment E, the typical effects of entanglement will be blurred. In the extreme case that the states $|E_i\rangle$ are mutually *orthogonal* (i.e. the states are without any overlap—this is the case if the environment responds completely differently to the various states $|\alpha\rangle_i \otimes |\beta\rangle_i$) the effects of entanglement will even become completely invisible in measurements on the composite system alone (i.e. if one does not look at E).[9]

It should be noted, however, that this disappearance of entanglement is not only approximate, but also relative to a limited class of observations. As inspection of Eq. 4.3 demonstrates, the total state of the original composite system plus its environment is still entangled—the process of decoherence has merely spread out the original entanglement so that it now also involves the environment E. As a consequence, observations of the original system *plus* the environment with which it has interacted will show the entangled nature of the total state, with its non-classical and non-local characteristics. However, it is true that if one restricts oneself to measurements on open systems *without* looking at their environments, *and* if one's measurements are not too precise, quantum effects will often[10] not manifest themselves and classical models of what happens will become possible. Another important consequence of decoherence is that open quantum systems in environments of the kind we are used to tend to become *localized*. This is because the usual interactions (electromagnetism, gravity) are sensitive to position, with the consequence that (practically) orthogonal environment states will become correlated to object states associated with different positions. By virtue of the same argument as before, superpositions of different positions will therefore become practically unobservable.

Summing up, quantum mechanics describes a world that is basically non-local and holistic, with properties of composite systems that generally do not supervene on the properties of their parts. But the process of decoherence is able to hide these typical quantum features from view. In particular, when we make observations on Earth, outside a fundamental physics laboratory, models on the basis of classical physics will normally work very well.

---

[9] Formally, the essential difference between the situations before and after interaction with the environment is that initially the expectation value of any operator O of the composite system alone is given by $\langle\Psi|O|\Psi\rangle = \sum_{i,j} c_i^* c_j \langle\alpha_i\beta_i|O|\alpha_j\beta_j\rangle$, whereas after the interaction with the environment this becomes $\sum_{i,j} c_i^* c_j \langle\alpha_i\beta_i|O|\alpha_j\beta_j\rangle\langle E_i|E_j\rangle$. The inner products $\langle E_i|E_j\rangle$ that have appeared tend to wash out the "cross terms", with $i \neq j$—these cross terms are needed to show the presence of entanglement. In the extreme case of orthogonality between different environment states we have $\langle E_i|E_j\rangle = 0$ if $i \neq j$, so that the effects of entanglement vanish completely from sight.

[10] In particular, in the circumstances of everyday observation. In laboratory experiments it turns out that quantum effects affecting even macroscopic objects can be made visible—these experiments on so-called Schrödinger cat states have become almost routine now.

So we may conclude that quantum mechanics leaves room for mechanistic explanations: there is a limited domain of quantum phenomena, defined by (a) a restriction on which parts of a total system are investigated, (b) a limitation on the accuracy with which these investigations are carried out, and (c) the presence of decoherence processes, in which mechanistic models apply. This seems in accordance with a conclusion recently drawn by Kuhlmann and Glennan, who write (Kuhlmann and Glennan 2014, 353)

> that decoherence provides a useful explanation of why, in particular local circumstances, systems behave classically in spite of their being ultimately constituted of entities that obey the principles of quantum mechanics, and that this explanation deflects possible concerns over the ontological and explanatory legitimacy of the mechanistic approach.

However, one should not overrate this result.[11] Although mechanistic models usually make extremely accurate predictions in familiar "classical" settings, taken completely literally these predictions are, even though very close, still wrong: a purely quantum mechanical calculation, taking into account entanglement and the non completely vanishing values of the factors $\langle E_i|E_j\rangle$ (see Footnote 9) will give other and, importantly, *better* predictions. So there are features of reality, detectable in principle, that show that the literal content of the ontological claims of the mechanistic explanation strategy is *false*.

The situation can be compared to others that we already encountered. Mechanical explanation by central forces, even though it was for some time the paradigm of mechanical explanation, turned out to be empirically inadequate when electrodynamic phenomena were investigated more extensively and in more detail. The law of Coulomb, according to which two electrical charges attract or repel each other by an inverse-square central force (in analogy to Newton's law of gravity) had to be replaced by a more complicated law for moving charges (this more complicated interaction can be derived from Maxwell's equations). Now, in many circumstances—in particular those that were familiar to researchers in the beginning of the nineteenth century—Coulomb's law still yields excellent predictions in spite of this complication. This is because the charges under investigation often do not move too fast—although they always move somewhat and are never perfectly stationary—and the deviations from Coulomb's law are minute anyway, hardly observable without ultra-sensitive experimental techniques. So there is a certain domain of electrodynamic phenomena that in spite of the validity of Maxwell's theory can be handled perfectly well, for all practical purposes, with the older Coulomb theory. Does this justify the conclusion that Maxwell's electrodynamics does not undermine the older ontological and explanatory claims? It seems clear that this is not the case. True, *explanations* by means of the Coulomb theory can often still be

[11] Kuhlmann and Glennan sometimes make statements that create the (what would be a mistaken) impression that there is absolutely nothing wrong with mechanistic explanations in semi-classical contexts, even given the validity of quantum mechanics; e.g. they say "In this paper we argue, in part by appeal to the theory of quantum decoherence, that the universal validity of quantum mechanics does not undermine neo-mechanistic ontological and explanatory claims as they occur within classical domains" (Kuhlmann and Glennan 2014, 337).

maintained after Maxwell, but these explanations depend for their ontological grounding on Maxwell's electrodynamics plus an argument that the new dynamical effects that occur (e.g., loss of energy by radiation) fall below the threshold of observational accuracy. Similar comments apply to many other examples from the practice of physics, in which explanations are still given on the basis of obsolete and false theories.[12] In all these cases the original ontological basis of the explanations *is* undermined, but this does not exclude that the explanations themselves, as argumentative patterns, are still useful.

4.5 Conclusion: Mechanisms, Explanations and Understanding in Physics

Even within classical mechanics the status of mechanistic explanation is not unchallenged. It is true that in the days when the theory was first proposed the new ideal according to which all physical processes should be explained as the result of pushes and pulls sparked off great enthusiasm, but this ideal had soon to be abandoned. The rules of mechanicism had to be stretched, first by admitting central action-at-a-distance forces, then by allowing the forces to become more complicated. The reason, of course, was the development of physical theory: theoretical schemes along Cartesian lines proved to be empirically inadequate, after some time central forces shared this fate, and not long thereafter the whole concept of action-at-a-distance forces became obsolete. In the meantime new mathematical frameworks had developed for the formulation of classical mechanics, like the Lagrangian an Hamiltonian formalisms, and these gave rise to very different patterns of explanation, for example via variational principles.

Still, even though the Cartesian push and pull paradigm has long been left behind as a fundamental and general scientific scheme, explanations along these lines remain useful. For example, even if we think that interactions between bodies are always mediated by fields (perhaps quantum fields) or complicated action-at-a-distance forces, it usually helps to visualize such interactions via the picture of Cartesian collisions—physics textbooks are full of pictures of this kind, even if the subject is quantum field theory. This is because simple mechanistic models, if they yield results that are not too far off the mark, provide us with qualitative understanding of a process: they enable us to see, without entering into detailed calculations, what the approximate outcome of a process will be. The familiarity of the push and pull scheme makes it intuitively manageable.

[12] It is sometimes argued that it is impossible to give valid explanations on the basis of false theories at all (see De Regt and Gijsbers 2016, and the volume of which that essay is a chapter, for recent discussions on this topic). We do not agree that it is impossible to explain without literal truth (see the next section)—but if this impossibility were to be accepted, this would clearly call the mechanist ideal into question in a more radical way than we do here.

The same comments apply to the other types of mechanistic explanation. For example, even in the context of general relativity it often helps to think of the gravitational attraction between material bodies in terms of the conceptual framework of Newton's theory. Why is it that light cannot escape from a black hole? Because the black hole is so massive that the gravitational force it exerts on light pulls the light towards the black hole so strongly that it cannot get away. Is is easy to understand the process this way, and it requires a lot more training to become equally familiar with the general relativistic scheme of null-geodesics, horizons and the Einstein field equations. Nevertheless, it is not impossible at all to acquire an intuitive familiarity with such advanced mathematical schemes; seasoned researchers do not need to make detailed calculations in order to make a qualitative judgment about, e.g., to what extent a solution of the field equations will deviate from Euclidean geometry, given a particular mass distribution.

The Lagrangian approach to classical mechanics, with its variational principle, illustrates our general point further. Although this approach pertains to cases in which Newtonian explanations that use forces are certainly also possible, there are circumstances in which one may nevertheless prefer an explanation along Lagrangian lines. This may happen if one wants to avoid anthropomorphic or indexical elements in one's explanations (see the discussion about the flow of time and four-dimensionality in Sect. 4.2), or if one wishes to lay stress on the continuity between classical mechanics and relativity theory; or on the analogies between mechanics and optics. Explanations via the Lagrangian framework are certainly not less ontologically grounded than their Newtonian causal counterparts: the structure of the four-dimensional world is such that it obeys variational principles—if anything, it is the Newtonian mechanistic explanation, with its *production* of effects during *the flow of time* that can be accused of introducing subjective elements. Moreover, in many cases one can develop an intuitive feeling for the outcome of variational arguments so that they make it possible to achieve understanding. For example, it is understandable why the trajectory of a free particle will be a straight line, as this path realizes the shortest distance. Also in cases with more complicated Lagrangians a similar geometric interpretation often makes it easy to make qualitative statements about the form of trajectories as shortest connection in some geometry.

We know from Sect. 4.3 that mechanistic explanations will be available in principle as soon as a Lagrangian scheme applies—for this, the Lagrangian does not even need to depend on mechanical quantities at all. What is more, there will be infinitely many different mechanistic explanations, of any sort one wishes: using contact forces, action at a distance, etc. It hardly needs argumentation, though, that this abundance does not help to enhance the attractiveness of such explanations: they will as a rule be too unwieldy, complicated and unnatural to be taken seriously. Clearly, the mere fact that the explanations in question are mechanistic does not compensate for this disadvantage—such "Poincaré schemes" are artificial and unenlightening, even though they are able to reproduce all empirical results correctly.

What this all points to is that explanation and understanding in physics are not restricted to one privileged standard format. There are usually several forms of explanation available, and which type is actually chosen in a particular situation depends on contextual factors like the exact question that is being asked (and its "contrast class"), the intended use of the explanation, and the conceptual framework that is adopted. The same applies to the notion of *understanding*. Since understanding is a more qualitative concept than explanation and also depends on factors like the skill of the actor who is involved and his/her familiarity with the theoretical framework, there is even more freedom here than in the case of explanation (De Regt and Gijsbers 2016; De Regt and Dieks 2005). In particular, it is not unusual to achieve understanding of physical processes with the help of obsolete theories that have been proven wrong when taken literally. Such theories may in spite of their incorrectness provide tools by which one can attain an intuitive grip on a process, and succeed in foreseeing its outcome in a qualitative way. This is particularly true for theories that use mechanical concepts. Quantum mechanics has supplanted these theories, but mechanical reasoning may still provide a conceptual grip on certain phenomena.

A final example may be useful here. In their plea that the validity of mechanistic explanation is not undermined by quantum mechanics, Kuhlmann and Glennan write (Kuhlmann and Glennan 2014, 357)

> why do flocks of birds so often form the inverted-l-shaped form often seen in autumn? A mechanistic explanation explains how this local phenomenon arises through the local interaction of the birds; global entanglement between the birds (and their constituents) and the rest of the universe are (to a high approximation) not causally or explanatorily relevant to the production of this phenomenon.

This is exactly right if construed as a proposal for one way of *understanding* how the shape of a flock of birds arises. But note that a mechanistic explanation in terms of productive forces is not the only possibility of achieving such understanding: a Lagrangian variational approach (e.g., in this case, in terms of finding a constellation of birds with maximum stability, by minimizing an energy expression) would work as well; it depends on contextual factors which approach is preferred.

Note further that the mechanical theory invoked in the explanation of the form of the birds flock is, taken literally, *wrong* (as acknowledged in the quotation by the addition of "to a high approximation"). Quantum theory is taken to be the more correct theory here, and it is in fundamental conflict with classical mechanics. It follows that a *better* explanation than the suggested mechanistic one is available if one is interested in the highest attainable predictive accuracy. It is true that the differences in cases like this will normally be astronomically small, but still in principle the suggested mechanical model will give results that are wrong in its details. The general statement that the quantum aspects of the situation are not causally or explanatorily relevant is therefore false. It depends on the *perspective* that is taken whether or not entanglement and other quantum aspects should be taken into account. From an already taken mechanical vantage point they do not play a role; from the perspective of quantum mechanics they *are* relevant. In particular, these very quantum aspects determine whether an approximate account in mechanistic terms will be viable at all; they are thus certainly relevant in an explanatory sense.

We therefore conclude that on the fundamental ontological level physics has moved away from mechanisms: quantum mechanics is fundamentally at odds with the image of composite systems whose properties are produced by the properties of their parts, via local interactions. In spite of this, for the purposes of explanation and understanding mechanistic reasoning remains an important conceptual tool. But it is not at all the only possible one: a toolkit of conceptual instruments is available, and it depends on contextual factors which one should be chosen.

References

Bell, J. 1964. On the Einstein Podolsky Rosen Paradox. *Physics* 1: 195–200.

Bohm, D. 1952. A Suggested Interpretation of the Quantum Theory in Terms of 'Hidden' Variables, I, II. *Physical Review* 85: 166–179/180–193

Darrigol, O. 2000. *Electrodynamics from Ampère to Einstein*. Oxford: Oxford University Press.

De Regt, H.W., and D. Dieks. 2005. A Contextual Approach to Scientific Understanding. *Synthese* 144: 137–170.

De Regt, H.W., and V. Gijsbers. 2016. How False Theories Can Yield Genuine Understanding. In *Explaining Understanding: New Perspectives from Epistemology and Philosophy of Science*, ed. S. Grimm, C. Baumberger, and S. Ammon, Chapter 3. New York: Routledge.

Dijksterhuis, E.J. 1961. *The Mechanization of the World Picture*. New York: Oxford University Press.

Glennan, S. 2002. Rethinking Mechanistic Explanation. *Philosophy of Science* 69: S342–S353. Supplement: Proceedings of the 2000 Biennial Meeting of the Philosophy of Science Association. Part II: Symposia Papers.

Kuhlmann, M., and S. Glennan. 2014. On the Relation between Quantum Mechanical and Neo-Mechanistic Ontologies and Explanatory Strategies. *European Journal for Philosophy of Science* 4: 337–359.

Maxwell, J.C. 1865. A Dynamical Theory of the Electromagnetic Field. *Philosophical Transactions of the Royal Society of London* 155: 459–512.

Maxwell, J.C. 1873. *A Treatise on Electricity and Magnetism*, Vols. I, II. Oxford: Clarendon Press.

Poincaré, H. 1890. *ÉElectricité et Optique*. Paris: G. Carreé.

Poincaré, H. 1902. *La Science et l'Hypothèse*. Paris: E. Flammarion. English translation (1905): *Science and Hypothesis*. London: Walter Scott Publishing Company.

Psillos, S. 2011. The Idea of Mechanism. In *Causality in the Sciences*, ed. P.M.K. Illari, F. Russo, and J. Williamson, 771–788. Oxford: Oxford University Press.

Salmon, W. 1984. *Scientific Explanation and the Causal Structure of the World*. Princeton: Princeton University Press.

von Neumann, J. 1932. *Mathematische Grundlagen der Quantenmechanik*. Berlin: Springer. English translation by Robert T. Beyer (1955): *Mathematical Foundations of Quantum Mechanics*. Princeton: Princeton University Press.

Chapter 5
Mechanistic Explanations Generalized: How Far Can We Go?

Brigitte Falkenburg

> *We have to change our mechanistic view from the crude atomism that recognizes only the motions of material particles in the void to a conception that admits such nonmaterial entities as fields, but for all of that, it is still a mechanistic world view. Materialism is untenable, but the mechanical philosophy, I believe, remains viable.*
>
> (Salmon 1984, 241)

Abstract My paper investigates the methodological continuity of mechanistic explanations from early modern science to current scientific practice, focusing on their generalizations in physics and beyond. Mechanistic explanations in early modern science draw on the analogy between processes in nature and the ways in which machines work, and this analogy has remained effective up to the present day. Today's machines rely on the advanced sciences ranging from quantum physics to computer science, just as the current generalizations of mechanistic explanations do. I will show how these generalizations fit in with a general mechanistic methodology of the "dissecting" sciences, from its origins in early modern science (in particular, Newton's analytic-synthetic method), to the constituent models of matter based on quantum theory, and the top-down and bottom-up approaches of the follower sciences of physics, such as current neuroscience. All these approaches have in common that they explain higher-level phenomena in terms of lower-level components and causes. I will discuss them under methodological and ontological aspects, compare some examples from current physical practice with the recent discussion on the "new mechanical philosophy", and sketch the scope and the limitations of a generalized mechanistic methodology.

I would like to thank Stuart Glennan for helpful comments and Joshua Rosaler for language corrections.

B. Falkenburg (✉)
Technische Universität Dortmund, Dortmund, Germany
e-mail: brigitte.falkenburg@tu-dortmund.de

B. Falkenburg, G. Schiemann (eds.), *Mechanistic Explanations in Physics and Beyond*, European Studies in Philosophy of Science 11,
https://doi.org/10.1007/978-3-030-10707-9_5

5.1 Introduction

In the last decades, philosophy of science began to take into account that the methods of empirical science are much more complex than the traditional philosophical views about deductive and inductive methods admitted. Given that there is no unified theory of physics, which may serve as unambiguous grounds of scientific explanation, the DN model of scientific explanation and several of its followers even fail for physics, the paradigm science. Nor is the current stage of theory formation in higher-level sciences such as biology or neuroscience promising with regard to the possibilities of explaining their models in terms of a fundamental theory. The recent philosophical focus on mechanistic explanations comes much closer to the practice of many disciplines, from physics to neuroscience. This focus, the so-called "new mechanistic philosophy", is associated with a shift from theoretical reduction to the explanation of the behaviour of complex systems in terms of the interactions between their parts, and to ontological reduction.

Mechanistic explanations may be a new topic in the philosophy of science, but they are very old. They trace back to early modern science, which in turn is rooted in ancient atomism (Popa 2018). In view of scientific change, it is striking that they are still in use from physics to neuroscience. How can mechanistic explanations still work, after the scientific revolutions of the twentieth century, which made the mechanistic worldview obsolete?

In the following, I give an answer to this question in three steps. *First*, due to the twentieth century scientific revolutions, mechanistic explanations obviously were subject to generalization. *Second*, mechanistic explanations in a narrow or generalized sense are an indispensable heuristic tool of the sciences. From early modern science to current scientific practice, a mechanistic methodology in a generalized sense has been at work. *Third*, this generalized mechanistic methodology is empirically most successful, given that it has an ontological counterpart in the layered structure of the physical world. Nature exhibits different levels of phenomena, from subatomic particles to atoms, molecules, and macroscopic bodies, including organisms, planets, stars, and galaxies.[1]

In Sect. 5.2, I explain the concept of a mechanism, introduce the distinction of one-level and multi-level mechanisms (Sect. 5.2.1) and sketch how to generalize mechanistic explanations (Sect. 5.2.2), based on the analogy between processes in nature and the ways in which machines work in current science and technology (Sect. 5.2.2.1). Recent philosophical accounts of mechanistic explanations, too, generalize them in view of the current sciences (Sect. 5.2.2).

In Sect. 5.3, I attempt to unravel the methodological and ontological aspects of mechanistic explanations. In order to do so, I will investigate the continuity of methods of the "dissecting" sciences (Sect. 5.3.1) from early modern science (Sect.

[1] To discuss the position of scientific realism adopted here is beyond the scope of this paper. However, the ontological reduction associated with mechanistic explanations has certain limitations (see Sect. 5.5).

5.3.1.1) to current scientific practice (Sect. 5.3.1.2), and sum up how these methods give rise to a general mechanistic methodology (Sect. 5.3.2). In order to clarify its ontological import, in (Sect. 5.3.3) I discuss ontological adequacy vs. theoretical truth (Sect. 5.3.3.1), dynamic part-whole relations (Sect. 5.3.3.2), the quantum parts of matter and light (Sect. 5.3.3.3), and the criteria for good mechanistic explanations, adding some tentative remarks on their scope and limitations (Sect. 5.3.4).

In Sect. 5.4, I finally discuss some typical examples of mechanistic explanations from current physics. Cases based on continuum mechanics (Sect. 5.4.1) and quantum theory (Sect. 5.4.2) shed some light on the current philosophical discussion, which seems to be too narrow in comparison to scientific practice. In Sect. 5.5, I will sum up my conclusions.

5.2 From Mechanisms to Mechanistic Explanations

The English word 'mechanism' goes back to the seventeenth century. Etymologically, it traces back to the Greek word μηχανή for 'machine' and the corresponding Late Latin word *mechanica* or its derivatives. According to the definition in a British dictionary, a mechanism primarily is

> 1. a system or structure of moving parts that performs some function, especially in a machine (Collins 2012).

According to this definition, mechanisms have a dynamic aspect, given that the function they perform is due to their moving parts. The proponents of the recent "mechanistic turn" of the philosophy of science also emphasize this dynamic aspect, but in quite different regards (see Schiemann 2019). They define the concept of mechanism in terms of causal processes or causal laws (Salmon 1984, 240; Glennan 1996, 52; see below Sect. 5.2.2.2), or in a more general way, e.g. in terms of causal activities (Craver 2007, 6).

Mechanistic explanations in the practice of science draw on the analogy between processes in nature and the mechanisms of machines. This analogy dates back to early modern science, and it is still effective today. A mechanistic explanation traces a phenomenon back to its components and/or causes in terms of the way in which a machine works or performs a function. The dictionary quoted above extends the definition of a mechanism to this analogical use and gives a secondary definition, according to which a mechanism is

> 2. something resembling a machine in the arrangement and working of its parts: the mechanism of the ear (Collins 2012).

This secondary, analogical meaning of the term 'mechanism' emerged in the seventeenth century, too, but it had precursors. Mechanistic explanations trace back to ancient atomism, which became most influential in early modern science, in the Cartesian mechanistic worldview as well as in the foundations of classical physics. Important mechanical explanations in early modern science relied on the analogy between the universe and a clock. Later, the laws of Newton's mechanics explained the structure of the moving parts of the celestial clock, or the machinery of the universe, in terms of gravitation as a universal force.

The extension of mechanical analogies to organs and their functions also is very old. Already Aristotle discussed the analogy between technical tools or machines and processes within nature (Aristotle, *Physics* 199a), albeit within his teleological account of nature. Early modern science dispensed with teleological explanations; hence, in seventeenth century mechanical philosophy, the analogy between mechanisms and the functions of organs is the other way round. Its ancient ancestor was atomism, in particular, the atomistic explanation of the secondary qualities of sensory perception. After the reception of the Arabian theory of vision in the Late Middle Age, and the rediscovery of atomism as described in Lucretius' *De rerum natura* (Lucretius 1675) in the Renaissance, mechanical (or technical) analogies entered medical science. The title of Andreas Vesalius' famous anatomy textbook *De humani corporis fabrica* (Vesalius 1543) paradigmatically expresses the analogy between the structure of the human body and an artificial structure. In the seventeenth century, Thomas Hobbes developed a most influential mechanistic theory of sensory perception (Hobbes 1655).

5.2.1 One-Level and Multi-level Mechanisms

Mechanisms, and the corresponding mechanistic explanations, may differ in complexity. In order to understand current mechanistic explanations and their precursors, it is important to make the following distinction. A *one-level mechanism* is a process restricted to the "horizontal dimension of spatio-temporal and causal organization" (Glennan 2017, 28). In it, the moving parts and the causal processes or activities of the machinery belong to the same level of components of a complex system as the function or effect brought about by them. A *multi-level mechanism*, on the other hand, is a mechanism operating within a complex system (*ibid.*). In it, the moving parts of the machinery bring about a function or effect at the higher level of the system as a whole. *One-level* and *multi-level* mechanisms are structurally different. The former explain the propagation of a causal process along the parts of a complex system. The latter explain how certain properties or activities of a complex system stem from the causal activities of its parts. They also have been distinguished as productive mechanisms and underlying mechanisms (Craver and Darden 2013, 66).

The above dictionary definition of a mechanism in terms of the process performed by the moving parts of a machine applies to one-level mechanisms, such as a lever or a pulley tackle, as well as to two-level mechanisms, such as a clock. A lever or a pulley tackle brings about the effect of raising a heavy body, by transferring the work directly at one-and-the-same level from the lever or pulling tackle to the raised body. A clock brings about the effect of moving the clock hand in order to indicate the time. It consists of gear wheels and a balance spring as moving parts, which move the clock hand at the surface of the clock. The moving parts belong to a lower level of the mechanism than the phenomenon of the time indicated at the surface of the clock.

A complex machine such as a car is multi-level. The car consists of non-moving parts such as the car body and the seats, and of moving parts such as the wheels, the axes, the motor and the steering wheel in order to perform the function of riding. The moving parts in turn consist of lower-level causal components that make them move, such as the spark plugs.

All mechanisms are complex systems. Whether a mechanism is considered as a one-level, two-level, or multi-level mechanism has epistemic and/or pragmatic aspects, depending on the focus of an explanation. To consider a clock as a two-level mechanism that indicates the time as a higher-level phenomenon has an epistemic aspect. In the case of a car, it depends on the pragmatic context that concerns which moving parts at which level are relevant, e.g., for the use or for the repair of a car. Given such conceptual ambiguities, Glennan (2017, 204) suggests to restrict the distinction of lower and higher levels to the "vertical" constitution of mechanisms, i.e., to the composition of wholes from parts. (Buzzoni 2019) discusses the perspectival aspects of the distinction.

Many mechanistic explanations of physics and other sciences aim at explaining the higher-level behaviour and the phenomenological properties of complex systems, and hence rely on multi-level mechanisms. Typical examples of physics concern the mechanisms of structure formation in the universe, from Descartes and Kant up to current astrophysics and cosmology. Descartes (1644) argued in favour of a vortex theory of gravitation, according to which the motions of the mechanical corpuscles of celestial matter exert the gravitational effects on the celestial bodies. The lower level of the respective mechanism consists in the corpuscles (as moving parts) and their impacts (as causal processes or activities). The higher level of the mechanism is the pressure their motions exert on the celestial bodies, which (according to Descartes) gives rise to the phenomenon of gravitation. Descartes' theory of light, however, employs the one-level mechanism of the propagation of impact between the corpuscles of celestial matter.

Kant, in his *Universal History and Theory of the Heavens* (Kant 1755), postulates a multi-level mechanism in order to explain the phenomena of astronomy and the observable structure of the universe. The mechanism of his theory of structure formation in the universe has more than four levels.

The lowest level consists in the atoms, endowed with attractive and repulsive forces as their causal activities. According to Kant, the atoms and their forces obey the laws of Newton's mechanics, making up the lowest level of the dynamic mechanism. The second level is a rotating matter nebula, within which (according to the causal processes of the atoms) celestial bodies agglomerate in a clustering process. The third level is the structure of the solar system that emerges from the system of rotating celestial bodies. Due to the law of gravitation (and an intuitive principle of conservation of angular momentum, which neither Newton nor Kant explicitly stated), the celestial bodies form the solar system as a system of approximately planar shape. The fourth level is the genesis of the Milky Way due to the same laws and principles, as an approximately planar system of stars like the sun (which we observe from a lateral position, in the solar system). On this level, relying on telescope observations of stellar nebulae and on Durham's theory of galaxies, Kant argues that the Milky Way is just one system of stars among many others. Finally, on a fifth level, Kant adds speculations about the coming into being and passing off of clustered star systems such as the Milky Way, in a speculative history of the evolution of the whole universe. His theory of structure formation in the universe even explains Kepler's paradox of how the night sky can be dark, in an infinite universe of infinitely many stars (Kepler 1610, De Bianchi 2013).

The multi-level mechanism behind Kant's mechanistic explanation of structure formation in the universe is dynamic. Its basis is the two-level mechanism of the clustering process in which celestial bodies form from a vortex of atoms. The next higher level concerns the formation of the solar system as a compound dynamic system of celestial bodies. Kant`s theory shows that in physics, a multi-level mechanism typically explains the formation of compound systems from the forces or interactions between their constituent parts.

5.2.2 Mechanistic Explanations Generalized

In view of the twentieth century revolutions of physics, the traditional mechanistic philosophy of Descartes, Newton, or Kant no longer works. The compound systems of physics at a small scale are no longer subject to classical mechanics but to quantum mechanics and quantum field theory, whereas the models of structure formation in the universe have to take into account the laws of special and general relativity. Hence, scientists as well as philosophers of science stated that the old mechanical philosophy had become obsolete. However, there also came suggestions how to generalize the mechanistic philosophy. An instructive example of both tendencies is due to William M. Malisoff, biochemist and lecturer in philosophy, member of the Advisory Committee of the *International Encyclopedia of Unified Science* and (from 1934 to 1944) member of the editorial board of *Philosophy of Science*. In the article *Physics: The Decline of Mechanism*, he claims that

> there has been a decline of mechanism, or if you will, an improvement on it. (Malisoff 1940, 401)

In order to support this claim, he sketches how relativistic and quantum physics dispensed with the assumption of a mechanistic ether and the knowledge of the complete set of initial conditions of the motions of subatomic particles (ibid., 408–411). Finally, he states:

> Do I think the decline of mechanism will be its death? Of course not. Mechanism still has and will have its uses. We merely see it only as a limited method. We merely see it as a partial view; we see that there is a more complete view which contains it as a factor, a limiting case. (ibid., 414)

5.2.2.1 Generalizations Based on Scientific Progress

Malisoff's conclusion concerning the decline of mechanism is its improvement by a more complete, materialistic view:

> The complete view I call materialistic. The decline of mechanism to me is the rise of materialism. Materialism is everything we have found matter capable of – and it is capable of being more than merely mechanical, even if it is but a poor shivering electron in an infinitely dull and cold field.

Hence, for him materialism based on the current scientific theories of matter is an improved version of the traditional mechanical philosophy. Malisoff does not directly suggest how to generalize the mechanistic view of nature on the lines of materialism, but he gives the crucial hints how to do so:

> What did the physicists of 70 years ago speculate about? I should say they speculated about mechanism itself. What is a mechanism? [...] A mechanism, they thought, is essentially a machine. And what is a machine? Simply enough, [...] a thing of cogs and levers. (Malisoff 1940, 405)

He emphasizes that the views of the classical physicists about the mechanisms of nature rely on a re-interpretation of the moving parts of mechanical machines in terms of idealized mathematical entities and the point masses and forces of mathematical physics:

> The difference, however, between a physicist and a machinist was that the physicist's cogs and levers and machines consisted of mathematical points, lines surfaces, volumes, interacting by a system of forces between the points to which were attributed masses and velocities. (ibid., 405-406)

Hence, it is obvious how to generalize the traditional mechanical physics to an up-to-date version (which Malisoff calls materialism), in the age of relativity and quantum theory:

> Do we not still use forces, particles, and the like, where we can? (ibid., 414)

A related way of generalizing mechanistic explanations is to invoke technological progress. seventeenth century mechanistic explanations draw on the analogy with mechanical machines. Descartes did so in his corpuscular theory of matter, vortex theory of gravitation and impact theory of light. Eighteenth/Nineteenth century mechanistic explanations relied on classical mechanics. Kant did so in his theory of structure formation in the universe, Maxwell and Boltzmann in the explanation of thermodynamic phenomena by the kinetic theory. In contrast, twentieth/twenty-first century mechanistic explanations may draw on the analogy with machines based on *any* kind of physical, chemical, or biochemical processes known by science. Hence, mechanisms in a generalized sense range from classical electro-dynamic circuits (which are the processes at work in the hardware of computers, analogously to those in the axons of nerves) to quantum mechanical processes (which are at work in the transistors of electronic devices).

The result is the same as suggested by Malisoff, namely an improved or generalized version of mechanistic explanations in terms of "forces, particles, and the like".

Current mechanistic explanations explain natural phenomena in terms of biological cell structures, cells, proteins, molecules, atoms, and subatomic particles; and they describe the causal activities of these entities in terms of artificial neural networks, signal transmission, electric currents and circuits, electromagnetic fields and waves, quantum processes and subatomic interactions.

5.2.2.2 Philosophical Generalizations

In the philosophy of science, Wesley Salmon emphasized that mechanistic explanations in a generalized sense are required for an adequate account of scientific explanation. The quest for a generalization of the traditional mechanistic philosophy is the same as in Malisoff's approach, with the following terminological difference. Malisoff generalizes his concept of materialism from the traditional mechanistic philosophy to explanations based on non-classical physics. Conversely, Salmon generalizes the concept of mechanistic explanation from classical to non-classical explanations, reserving the concept of materialism to the traditional mechanistic world view. In his book *Scientific Explanation and the Causal Structure of the World*, he suggests an ontic conception of scientific explanation in terms of causal mechanisms:

> [...] what constitutes adequate explanation depends crucially upon the mechanisms that operate in our world. (Salmon 1984, 240)

He emphasizes that mechanistic explanations may even employ fields (ibid., 241). According to him, a mechanism is any causal fork or causal process, including stochastic processes:

> The theory here proposed appeals to causal forks and causal processes; these are, if I am right, the mechanisms of causal production and causal propagation that operate in our universe. These mechanisms [...] may operate in ineluctably stochastic ways. (ibid., 239)

Hence, according to Salmon mechanistic explanations are causal explanations. At first he defined causal processes in terms of mark transmission (1984), and later, in terms of the transmission of a conserved quantity between two events (1998). According to both definitions, the paradigm case of a causal process is signal transmission in physics,[2] such as the emission, propagation and detection of radio waves or light signals, including quantum processes such as the emission, propagation and absorption of photons.

Salmon's 1984 conception of a mechanism is very general. It holds for the classical impact of two billiard balls as well as for the transmission of a quantum signal, which obeys the principle of energy conservation and the probabilistic laws of a quantum theory. In addition, it is very basic. His account of mechanisms and his

---

[2] Both definitions fall short of a *general* concept of causality, as shown by the everyday example of causation by omission (Dowe 2008). However, Salmon's approach is concerned with scientific explanations. It deals with natural causal processes, not with everyday explanations of human actions.

paradigm case of signal transmission concern one-level mechanisms. Signal transmission is a causal process that propagates from cause A to effect B, where A and B belong to one-and-the same level of phenomena.

In contradistinction to Salmon's approach, the mechanisms discussed in the recent philosophy of science are multi-level. In accordance with Stuart S. Glennan's definition of a mechanism, they concern complex systems and their causal components:

> A mechanism underlying a behavior is a complex system which produces that behavior by […] the interaction of a number of parts according to direct causal laws. (Glennan 1996, 52)

Glennan's approach applies to multi-level mechanisms in physics.[3] It expresses the causal processes of the components of a mechanism in terms of interactions that obey direct causal laws. His goal is to explain higher-level causal processes in terms of physical mechanisms, whereas the fundamental laws of physics, which serve to explain the higher-level processes, according to him are not subject to mechanistic explanation.

Kant's theory of structure formation in the universe (see above Sect. 5.2.1) is in perfect agreement with Glennan's definition. Here, the laws of Newton's mechanics, in particular, the law of gravitation, are the "direct causal laws" which give rise to the agglomeration of bodies, the formation of the solar system and higher-level systems of celestial bodies.

In the current practice of physics, it is quite often possible to express the causal processes underlying a mechanism in the precise terms of the laws of a physical dynamics (see Sect. 5.4). In computer science, it is possible to express them in terms of mathematical algorithms.

In biology, this approach is possible to the extent that chemical or biochemical mechanisms are at work, which in turn reduce to the mechanisms of molecular physics; that is, in biophysics, genetics and biotechnology. A good example of applying the laws of physics in a mechanistic explanation to biophysical processes is the computer simulation of protein folding. In cell biology, epigenetics, evolution theory, neurobiology, etc., the situation is much more complex and hence more difficult. Here, in many cases the causal laws available in order to explain the way in which the causal components of a mechanism work are laws in a weak sense, as suggested in (Glennan 1996, 2002); or there are no known causal laws at all, as in the case of the heuristic assumption of mental mechanisms (Bechtel 2008). Therefore, philosophers of biology and neuroscience typically define a mechanism in very general terms, as in the seminal work of Machamer, Darden and Craver:

[3] Glennan considers his approach to be more adequate than Salmon's. In my view, they just concern two different kinds of mechanisms, namely multi-level versus one-level mechanisms. Current scientific practice employs both; see my examples discussed in Sect. 5.4. – Schiemann (2019) points out another difference neglected here, namely that of ontological monism vs. dualism. Salmon's analysis of causal mechanisms in terms of the transmission of conserved quantities between events is dualistic.

> Mechanisms are entities and activities organized such that they are productive of regular changes from start or set-up to finish or termination conditions. [...] To give a description of a mechanism for a phenomenon is to explain that phenomenon, i.e., to explain how it was produced. Activities are the causal components in mechanisms. Mechanisms are composed of both entities (with their properties) and activities. Activities are the producers of change. Entities are the things that engage in activities. Activities usually require that entities have specific types of properties. (Machamer et al. 2000, 3)

Similarly, Craver defines:

> Activities are the causal components in mechanisms. [...] mechanisms are entities and activities organized such that they exhibit the explanandum phenomenon. (Craver 2007, 6)

The definition of Bechtel and Abrahamson is even more general:

> A mechanism is a structure performing a function in virtue of its component parts, component operations, and their organization. The orchestrated functioning of the mechanism is responsible for one or more phenomena. (Bechtel and Abrahamsen 2005, 423)

Even though the term "machine" does not appear in these definitions, their similarity to the dictionary definitions given above (Sect. 5.2) is striking. According to them, a biological or neural mechanism works like a machine. Mechanisms in this sense explain higher-level phenomena in terms of lower-level causal components, in a "dualistic" account of the components and their causal activities (Schiemann 2019). Such explanations do not clarify which kinds of causal activities are at work, and how they relate to the causally active entities. They just rely on the analogy between a complex system in nature and a machine with well-defined moving parts.

In particular, in cognitive neuroscience, the explanation of mental phenomena in terms of neural mechanisms draws on the analogy between the brain and a parallel computer, i.e., an artificial neural network. The way in which the causal activities of the neural network in the brain give rise to cognitive capacities is to some extent known, but the way in which they give rise to mental phenomena is completely unknown.

5.3 Methodological and Ontological Aspects of Mechanistic Explanations

The scientific and philosophical generalizations of mechanistic explanations sketched above are not arbitrary, as I will demonstrate now. Mechanistic explanations have substantial ontological import. They rely on the ontological assumption that nature has a layered structure that the machine analogy behind mechanistic explanations captures to a certain extent. The machine analogy behind mechanistic explanations is an idealized model of great heuristic value. In early modern science, it laid the foundations for a most fruitful heuristics that became empirically successful in the nineteenth century and made the methodological bridge to twenty-first century physics, biochemistry, neurobiology, and computer science.

Methodologically, mechanistic explanations rely on a significant, though little investigated, continuity of the scientific methods from early modern science to the present day. The core of the mechanistic methodology consists in dissecting the phenomena into components and studying the way in which the causal activities, or interactions, of the latter give rise to the former (as claimed in Bechtel and Richardson 2010). The ontological import of mechanistic explanations is due to the fact that the layered structure of nature and the composition of complex systems to a certain extent permits this.

The practice of dissecting the phenomena gave rise to a general mechanistic methodology, which has been at work from early modern science, and in particular, in Newton's analytic-synthetic method, until today, in the *top-down* and *bottom-up* approaches of current scientific practice. The generalized mechanistic methodology attempts to dissect complex systems *top-down* into their components and to investigate the causal activities of the latter. Vice versa, it aims at explaining the behaviour or the dynamic properties of the complex systems *bottom-up* from the causal components. In twentieth century science, this generalized mechanistic methodology bore its fruits, at the price that the theory underlying the traditional mechanical philosophy, classical mechanics, turned out to be wrong. Hence, one may say that in order to become successful, the mechanical philosophy had to throw its classical mechanical restrictions overboard. Its main empirical successes are the ontological reduction of bodies, fluids, or gases to molecules, atoms, subatomic particles, and their interactions; of organisms to organs, cell structures, cells, cell nuclei, DNA, proteins, biochemical substances, and chemical processes; of the brain to brain areas, neural networks, neurons, axons and dendrites, and biochemical neurotransmitters. To discover these entities and their causal activities or interactions was beyond the scope of early modern science, but Newton's analytic-synthetic method laid the grounds.

5.3.1 The "Dissecting" Sciences

The method of dissecting the phenomena is the most important distinguishing mark of the sciences of the modern age. In spite of all scientific change, from Galileo's and Newton's days up to current scientific practice, the natural sciences have this methodological feature in common. Physics, chemistry, biochemistry, biology, or neuroscience all aim at the decomposition and re-composition of their respective phenomena. In contradistinction to ancient astronomy and natural philosophy, early modern science did so not just by means of conceptual and mathematical analysis, but also by means of systematic experiments. The experiments of the empirical sciences aim at decomposing and re-composing the phenomena in order to find their causes. It is for this reason that they are called the "dissecting" sciences (Schurz 2014, 35). It is not by chance that this expression calls to mind the practice of medicine and biology of dissecting human bodies, animals, and plants. The practice of dissecting the bodies of animals and human beings laid the grounds for early modern science. In the age of the Renaissance, it gave rise to the extension of mechanical analogies to organs and their functions. In particular, the title of Vesalius' anatomy textbook *De humani corporis fabrica* (Vesalius 1543) already mentioned above (at the end of Sect. 5.2) indicated the idea of fabrication, and hence, a mechanistic view of the human body. Galileo transferred the practice of dissecting the phenomena from medicine to the experiments of physics.

As a dissecting science, physics decomposes physical objects and phenomena in the broadest sense into their parts or components. The experiments of physics decompose bodies, fluids, and gases, but all kinds of radiation, too, including light, as in Newton's optical experiments. They do so in order to explain their phenomenological properties and behaviour in terms of causal components. That is, they aim at mechanistic explanations; and so do the follower sciences of physics, in a remarkable continuity of scientific methods from early modern science to the present day.

Indeed, the dissecting sciences and the underlying mechanical analogies in a broadest sense dominate our current scientific worldview. Given the enormous impact of the dissecting sciences on our scientific worldview, it is hard to understand why they have not yet drawn the attention they deserve in philosophy of science.[4]

5.3.1.1 The Traditional Analytic-Synthetic Method

The founders of early modern science depicted the methodological aspects of mechanistic explanations, or the dissecting methods they used, in terms resolution and composition (Galileo) or of analysis and synthesis (Newton). The method dates back to ancient geometry (Pappus [1589] 1986). According to Pappus, analysis and synthesis are the complementary parts of a combined regressive-progressive method. Its "analytic" part is regressive or inductive. It proceeds from something that is given or assumed to the underlying principles. Its "synthetic" part is progressive or deductive. It aims at confirming the principles by deriving from them the original data or assumptions. According to Pappus, this was a mathematical method. In early modern science, analysis and synthesis transformed respectively into the decomposition of phenomena and their re-composition from their parts and causes.[5]

For Galileo and Newton, analysis and synthesis, or (in Galileo's terms) the resolutive-compositive method, had two aspects, one related to experimental and mathematical decomposition of the phenomena, the other to causal analysis. For both physicists, the inductive part of the method (*analysis* or *resolutio*) was the regression from the phenomena, as the given data, to their components and causes, as the principles sought-after. This regression from what is given or assumed to these principles relied on experiments and observations. It was exploratory, or heuristic. The deductive part of the method (*synthesis* or *compositio*) progressed in the opposite direction, from the parts and the causes to the phenomena. It was

[4] Even the book from which I adopted their name here (Schurz 2014, 35-36) sketches their significance only in a very short paragraph.

[5] For more details concerning the method, its origins, and its transformations in early modern science, see Beaney 2015.

explanatory, and it had the structure of a mechanistic explanation. In *Query 31* of the *Opticks*, Newton describes the method as follows:

> By this way of Analysis we may proceed from Compounds to Ingredients, from Effects to their Causes, and from Motions to the Forces producing them; and in general, from Effects to their Causes, and from particular Causes to more general ones, till the Argument end in the most general. This is the Method of Analysis: And the Synthesis in assuming the Causes discover'd, and establish'd as Principles, and by them explaining the Phaenomena proceeding from them, and proving the Explanations. (Newton [1730] 1979, 404-405)

The quotation shows that for Newton the causal analysis of the phenomena has two aspects; first, the search for the ingredients or parts of a given compound; and second, the search for the forces behind the motions. Hence, Newton had a dualistic ontology in Schiemann's sense (see chapter 3 of the present volume).

In the *Principia,* Newton spoke of "induction" or "deduction from the phenomena".[6] Roger Cotes, editor of the second edition of the *Principia*, however, took the method of analysis and synthesis up in his preface in order to explain the scientific method of the founders of empirical science. His words are very close to Newton's remarks in *Query 31* of the *Opticks* quoted above:

> [...] they proceed by a twofold method, analytic and synthetic. From certain selected phenomena they deduce by analysis the forces of nature and the simpler laws of these forces, from which they then give the constitution of the rest of the phenomena by synthesis. (Newton [1713] 1999, 386)

In the *Principia*, Newton explains the motions of the celestial bodies and other mechanical phenomena in terms of gravitation as a universal force. In the *Opticks*, he explains the constitution of light and other optical phenomena in terms of colours and hypothetical light atoms. In particular, he shows how a prism decomposes white light into the colours, and how a second, parallel prism recomposes white light by superposing the spectra from both prisms (Newton [1730] 1979, 147).

In both fields of physics, the phenomena are the starting point of causal analysis. The crucial difference between the *Principia* and the *Opticks* is that Newton is able to give a full-fledged mathematical theory of gravitation explaining the phenomena of mechanics, whereas he is *not* able to do so for the atomic constituents of light or matter and their forces. Without establishing well-defined, empirically confirmed principles that explain the phenomena under investigation, the second step of the analytic-synthetic method, the synthetic part, is missing, and the method remains in his view incomplete. We may say that in order to give a well-established mechanistic explanation, in Newton's view the ontological reduction of a whole to its parts and causes must come together with a theoretical reduction of the whole to its compound parts and their interactions. The theoretical reduction is the deduction of the behaviour of the whole from a theory of its parts (and hence belongs to the synthesis part of the explanation). It explains the whole as a compound system (such as the solar system made up of the sun, planets, and moons, or an atom, as an N-particle

[6] For its relation to Newton's methodology of the *Principia,* as depicted in the famous *Rules of Reasoning* ((Newton [1713] 1999, 795 pp.), see Falkenburg 2017.

system of electrons in bound states in the field of the atomic nucleus) in terms of a dynamics of its parts.

Newton's *Opticks* gives a rough mechanistic explanation of the composition of white light in terms of light atoms of different colors, but it does not include precise knowledge of the causal processes and laws to which the emergence of white light from the colored light atoms is due. By contrast, Newton's *Principia* do explain the mechanism of the motions of the bodies in the solar system in precise mathematical terms, by the law of gravitation. In the mechanistic explanations of classical mechanics, ontological reduction comes together with theoretical reduction based on the dynamics of a complex system. In Newton's mechanistic explanation of light, it does not.

5.3.1.2 Current Top-Down and Bottom-Up Methods

Newton's rough mechanistic explanation of the composition of white light in the *Opticks* is similar to the current situation in biology and neuroscience. In neuroscience, there are rough mechanistic explanations of cognitive capacities and mental phenomena in terms of neural mechanisms, but they do not include precise knowledge of the causal processes and laws due to which cognitive capacities and mental phenomena emerge.

In current neuroscience, the terminology of *top-down* and *bottom-up* approaches replaced the traditional terminology of analysis and synthesis. Galileo's *resolutio* or Newton's *analysis* is the experimental decomposition of phenomena and the regression to their parts, causes, and (if possible) fundamental laws of a physical dynamics. Galileo's *composition* or Newton's *synthesis* is the experimental re-composition and the mechanistic explanation of phenomena from their parts, causes, and (if possible) fundamental laws of physics.

Obviously, Newton's *analysis* proceeds *top-down* from a mechanism in the above sense to its lower-level components and their causal activities, or forces; and Newton's *synthesis* proceeds *bottom-up* from the compound parts and their forces to an explanation of the higher-level phenomenon. In an analogous way, the *top-down* approaches of current biology proceed from organisms to organs and cells, down to cell structures and the causal activities of the genes made up from proteins and their biochemical components. The *top-down* approaches of neurobiology proceed from the brain to brain areas, down to neurons, axons, dendrites, synapses, and the electrochemical activities inside the axons and the biochemical effects of the neurotransmitters transmitting signals at the synapses. Conversely, the development and the functions of organs are explained *bottom-up* via the mechanisms of DNA reduplication, gene expression, etc.; and the cognitive capacities of the brain such as learning, memory, etc., are as far as possible explained in terms of neural mechanisms.

So far, the *top-down* and *bottom-up* approaches of current biology and neuroscience correspond to the complementary parts of Newton's analytic-synthetic method. The *top-down* approach proceeds from higher-level phenomena to lower-level structures. It gives rise to an inference to the best explanation in terms of causes.

Vice versa, the *bottom-up* approach proceeds from lower-level causal components and their activities to their higher-level effects, giving rise to mechanistic explanations. Just as in the case of Newton's *Opticks*, here a full-fledged theoretical reduction of the whole to its compound parts is still missing.

5.3.2 The Generalized Mechanistic Methodology

Thus, Galileo's resolutive-compositive method or Newton's analytic-synthetic method and the *top-down* and *bottom-up* approaches of current biology and neuroscience share the crucial structural features of their methods. They have in common to decompose natural phenomena into lower-level causal components in order to give mechanistic explanations. To dissect the phenomena in this way in order to explain them in terms of mechanisms is the general mechanical methodology of early modern science that finally became successful in nineteenth/twentieth century science. It draws on the analogy between processes of nature and the way in which machines work. In comparison to the traditional mechanical philosophy, it is more general in two regards. First, it is no longer restricted to classical mechanics. Second, it gives rise to ontological reduction without need of a full-fledged theoretical reduction.

5.3.3 The Adequacy of Mechanistic Explanations

This point needs further analysis. Let us look at the way in which classical mechanics brings theoretical reduction and mechanistic explanation together. The law of gravitation gives rise to a dynamics that describes a compound system of bodies moving around their centre of gravity. The compound system is the solar system; the celestial bodies are its components; their causal activities are the gravitational interactions. The law of gravitation is the dynamic law of a fundamental theory, or a "direct causal law" in Glennan's sense (Glennan 1996).

Many of the heuristically most fruitful mechanistic explanations rely on a *top-down* analysis of the phenomena, without giving a corresponding complementary *bottom-up* explanation in terms of such dynamic laws.[7] In many cases, the respective dynamics is just unknown. This was not only true of Newton's atomistic theory of light, but also of Kant's theory of structure formation in the universe. It could not yet rely on a precise atomistic theory of matter in order to explain the astronomic higher-level phenomena in terms of a physical dynamics. Mechanistic explanations may be good heuristic tools that capture *some* crucial features of the layered struc-

[7] I leave the question of fundamentality aside here, focusing on the marks of a good mechanistic explanation.

ture of nature, and thus to a certain extent be ontologically adequate, without relying on a full-fledged dynamics of the respective mechanism.

The condition to describe a complex system in terms of the dynamics of a compound system of lower-level parts, however, is also in another regard too strong. Today, the dynamics of light 'atoms' unavailable to Newton is expressed in terms of quantum electrodynamics, an empirically most successful theory. However, it does not describe light as a compound system of light quanta, but as a free quantized field. Hence, even if a well-established physical dynamics exists, the above condition seems too restrictive.

These examples show that the existence of a full-fledged dynamical description of a complex system is in general not necessary for a good, adequate mechanistic explanation. What then are the criteria? If the theoretical reduction of a whole to its causal components in terms of the dynamics of a compound system is not necessary, is it sufficient?

5.3.3.1 Ontological Adequacy vs. Theoretical Truth

At this point, we have to distinguish the ontological adequacy of a mechanistic explanation and the truth of the underlying theory. The kinetic theory of heat is a paradigm case of a good mechanistic explanation, but it is not true. It explains many higher-level phenomena of thermodynamics in terms of the kinetic energy of the atoms or molecules as the lower-level components of a gas. The corresponding physical dynamics is classical statistical mechanics, which describes the elastic collisions of the particles that make up a gas. By adding an appropriate initial condition (the assumption of molecular chaos), Boltzmann was able to derive a statistical explanation of the second law of thermodynamics from it, including the statistical increase of entropy. However, not even the founders of kinetic theory, Maxwell and Boltzmann, believed in the truth of their mechanistic explanation. They just thought that the kinetic theory is an idealized model (or picture, in Hertz' sense) of physical reality, but doubted to what extent it is ontologically adequate. Today, quantum theory indeed shows that the classical kinetic theory gives an inadequate description of the atoms and molecules of a gas.

Nevertheless, the theory was heuristically most fruitful. It gives a mechanistic explanation of the second law of thermodynamics, which is adequate with regard to the assumption that a gas consists of atoms or molecules with a certain distribution of kinetic energy, as the causal components to which the phenomena of thermodynamics are due. Hence, ontological reduction is possible without theoretical reduction, and a mechanistic explanation may be good and heuristically fruitful without being ontologically adequate in a strict sense. The mechanistic explanation of matter and the laws of thermodynamics in terms of classical corpuscles or particles turned out to be wrong. For its heuristic success it was sufficient that it was only ontologically adequate to a certain extent, with regard to the features of physical reality it indeed captures.

Mechanistic explanations rely on the machine model of processes in nature. Hence, just as other models, they are theoretical instruments that need not to be true in order to be explanatorily successful (see Morgan and Morrison 1999). The case of kinetic theory shows that the necessary and sufficient conditions for a good mechanistic explanation are weaker than full ontological adequacy (not to speak of truth), including the theoretical reduction of the whole to its causal components and their dynamics. Here, "good" means ontologically adequate just to a certain extent, and for this reason, heuristically fruitful for the explanatory goals of science. What, then, are these weaker conditions?

5.3.3.2 Causal Parts and Their Dynamic Properties

Much philosophical confusion about the legitimacy of generalized mechanistic explanations arose from taking the mechanical analogy with the moving parts of machines too literally. The parts of a mechanism in the generalized sense proposed here, however, do not need to be local, spatially well-identified parts.[8]

Why? The answer derives from the very concept of a mechanism employed in mechanistic explanations. To talk of the causal components of a mechanism implies two kinds of relation: a part-whole relation, and a causal relation. Given that a mechanistic explanation explains the behaviour or functioning of a mechanism in terms of its causal parts, however, a mere spatial interpretation of the part-whole relation of a mechanism does not match the way in which the moving parts of the machine function. The relevant part-whole relation concerns the causal properties, not the spatio-temporal properties, of the mechanism as a whole and its components. It is a relation between the causal or dynamic properties of the whole, on the one hand, and of its parts on the other. The parts of a mechanism act as dynamic parts. Their causal activities correspond to dynamic properties and the concepts expressing them. From a philosophical point of view, the expressions "causal parts" and "dynamic properties" are unclear.[9] In view of scientific practice, physics-based approaches to causal laws and processes should be preferred, as Glennan (1996) and Salmon (1984) suggested.

A difficulty of employing the causal laws and processes of physics is that different physical laws and theories are associated with several accounts of causality, from Einstein causality - that is, the deterministic transmission of a physical signal within the light cone - to the irreversible processes that cause an entropy increase, or the undetermined effects of a quantum measurement. Up to now, there is no

[8] This claim holds for the generalizations sketched in Sect. 5.2.2 as well as in Sects. 5.3.1, and 5.3.2.

[9] From a philosophical point of view, the expressions "causal parts" and "dynamic properties" are unclear. Current philosophy spells dynamic properties out in terms of dispositions, whereas the concepts of causality range from current successors of Hume's regularity theory over variants of Salmon's physics-based approach to Woodward's interventionist account, not to speak of Kant's a priori. Here, I leave the philosophical discussions on dispositions and causality aside.

unambiguous, well-established concept of causality or theory of causal processes that the physics community would share. Given that there is no unified theory of physics, the diverse concepts of causality used in the context of different theories cannot be unified neither.

Nevertheless, it is possible to generalize the notion of a mechanism in an uncontroversial way, in view of the practice of physics. At this point, we may recall Malisoff's observation that the physicists have always interpreted the machine account of mechanisms in terms of idealized mathematical entities, in such a way that

> [...] the physicist's cogs and levers and machines consisted of mathematical points, lines surfaces, volumes, interacting by a system of forces between the points to which were attributed masses and velocities. (Malisoff 1940, 405–406)

He suggested how this idealization makes it possible to improve (or generalize) the account of a mechanism in the age of relativity and quantum theory (see Sect. 5.2.2.1):

> Do we not still use forces, particles, and the like, where we can? (Malisoff 1940, 414)

Beyond the crude model of mechanical machines, the spatio-temporal structure of the parts of a mechanism is not decisive for the way in which it works. As Malisoff observed, this even holds for the mechanism of the solar system. Classical mechanics replaces the celestial bodies by point masses, given that their extension is negligible as compared with the distance between them. It describes the causal properties of physical systems in terms of dynamic magnitudes such as mass or charge. On this basis, beyond classical mechanics *any* physical dynamics may give rise to mechanisms, from electrodynamics to thermodynamics to quantum mechanics, quantum field theory, or general relativity.

Any physical dynamics, then, expresses the causal part-whole structure of a mechanism in terms of the dynamic properties of the whole and its constituent parts. According to physics (from Newton to current physical practice), the dynamic properties of physical systems and their components are physical quantities such as mass, charge, energy, and so on. Particles in a generalized sense are nothing but collections of such dynamic properties, for which conservation laws hold. This generalized particle concept reminds of Hume's bundle theory of substances, but it is stronger (or employs a causal aspect) with regard to the conservation laws of physics. The concept of a particle as a collection of conserved quantities is what remains in quantum physics of the classical particle concept, given that the quantum revolution dispensed with particle trajectories. Empirically, it corresponds to an operational particle concept, according to which a particle is the amount of energy, charge, etc. absorbed by a particle detector (Falkenburg 2007, chapter 6.3). Theoretically, it corresponds to Wigner's famous definition, according to which particles (or fields) are the irreducible representations of symmetry groups (Wigner 1939). According to this most general particle concept, the relation between particles and forces, or interactions, rests on dynamic symmetries that are associated with conservation laws for mass-energy, charge, spin, parity, and so on.

5.3.3.3 The Quantum Parts of Matter and Light

In quantum physics, the relation between the generalized particle concept in Wigner's sense, the conservation laws of physics, and the symmetries of a physical dynamics gives rise to sum rules that hold for a complex system and its parts. Such sum rules underlie the constituent models of current atomic, nuclear, and particle physics, and they are empirically testable. They hold for the mass and charge of an atom, and the electrons, protons, and neutrons of which it consists; here, the binding energy of the protons and neutrons has to be included in the sum of the conserved dynamic quantities mass and energy. They hold for the mass, charge, or spin of the quarks within the proton and neutron. There are specific high energy scattering experiments of particle physics designed to test the corresponding sum rules (see Falkenburg 2007, chapters 4 and 6.5). They hold for the number and kinds of quasi-particles of a solid investigated in condensed matter physics (Falkenburg 2015) as well as the strength of an electromagnetic field and the occupation number of the corresponding quantum field; or, for the intensity of light and the expectation value of the number of photons in it.

In all these examples, the quantum parts of matter and light are not subject to a spatio-temporal, but to a dynamic part-whole relation. The mechanistic explanations of current quantum physics rely on sum rules that hold for conserved quantities and the corresponding quantum dynamics, as far as known. To establish mechanistic explanations in the quantum domain, it is sufficient to know the sum, based on measurement and theoretical knowledge, and to test which sum rules and conservation laws for which quantities hold. Hence, contrary to the claims of Kuhlmann and Glennan (2014), in order to give support to neo-mechanistic ontologies and explanatory strategies for quantum systems, it is not necessary to employ the decoherence approach.

In quantum physics, to establish a sum rule for conserved dynamic quantities, based on measurement and supported by some kind of physical dynamics, seems to be a necessary and sufficient condition for a good mechanistic explanation. Such a generalized mechanistic explanation is very often two-level. It explains a complex system in terms of some dynamic parts, and it supports the ontological claim that the respective system consists of such dynamic parts. This condition is weaker than a full-fledged theoretical reduction in terms of the many-particle dynamics of a compound system. However, it also applies to Salmon's 1984 account of mechanistic explanations based on one-level mechanisms, at least as far as causal processes are identified with the transmission of a conserved quantity (Salmon 1998). In the last analysis, the transmission of a conserved quantity is the propagation of a signal in a complex system, according to the rules of some quantum dynamics.

5.3.4 What, Then, in General, Is a Good Mechanistic Explanation?

The above result also sheds light on the general question of what is a good, ontologically roughly adequate, mechanistic explanation (in the generalized sense proposed here). A good mechanistic explanation seems to rely on empirically justified dynamic part-whole relations. A sufficient condition seems to be that a whole can be shown to have dynamic properties that reduce to corresponding dynamic properties of some of its parts, where the dynamic properties of the parts sum up to the corresponding properties of the whole.

To establish sum rules of this kind stood at the beginnings of modern chemistry, the kinetic theory, and atomic physics, in the nineteenth century. Chemistry proceeded by establishing the law of multiple proportions, supporting the theory that chemical substances are composed of chemical elements, where the former are made up of molecules and the latter of atoms, and molecules are composed of atoms. The beginnings of modern atomic physics were due to measuring the elementary electric charge in the experiments of electrolysis. The kinetic theory proceeded due to Boltzmann's statistical explanation of entropy, which gave further support to atomism. All these developments advanced ontological reduction, either without being supported by theoretical reduction, or relying on incorrect theoretical explanations.

Is the above sufficient condition concerning dynamic properties and the corresponding sum rules for conserved quantities necessary, too? The mechanistic explanations of cognitive neuroscience suggest an answer in the negative. Mental phenomena and the cognitive capacities of human beings and animals, on the one hand, and the neural mechanisms in the brain, on the other hand, do not have any obvious common dynamic properties, for which it would make sense to attempt to establish sum rules and conservation laws, like in physics. The brain and the mind, the neurons and our ideas, do not stand in dynamic part-whole relations. They just are both localized in our heads, and it is possible to find and investigate spatio-temporal correlations between them, i.e., between neural activities in certain brain areas, on the one hand, and certain contents of a test person's consciousness or certain recognition capacities of a human being or an animal. In his book *Mental Mechanisms*, William Bechtel emphasizes that cognitive neuroscience even may employ heuristic identity assumptions about mental phenomena and their physical basis in neural mechanisms:

> One of the virtues of viewing identity as a heuristic claim is that it can guide not only the elaboration of the two perspectives which are linked by the identity claim, but it can use each to revise the other. (Bechtel 2008, 71)

The explanation of cognitive capacities in terms of neural mechanisms differs in a crucial regard from mechanistic explanations in physics, chemistry, or biochemistry. It makes use of an analogy between artificial and natural neural networks and of a concept of information based on computer science. Even in a very general sense of 'mechanism', this does not seem to be a mechanistic explanation proper. Nevertheless, the mechanistic explanations in terms of neural mechanisms are heuristically most fruitful. Hence, it is most unclear how far we may go in generalizing mechanistic explanations. Concerning the scope and limitations of mechanistic explanations, we should just keep two issues in mind. First, they draw on the machine analogy of natural processes, which is just a crude, idealized model. Like all models, they may be heuristically most fruitful, though strikingly wrong. Second, the "dissecting" sciences might have a limited scope concerning the investigation of the human mind.

5.4 Mechanistic Explanations in Current Physics

Let me conclude by giving some typical examples of mechanisms from the current practice of physics, which may shed light on the philosophical discussion of mechanistic explanations and their generalization. They stem from several areas of physics, including cases of applied physics that stem from neighbouring disciplines such as streaming mechanics or marine geology, and concern very different kinds of phenomena and mechanisms.

An internet search combining the terms "mechanism" and "physics" resulted in finding physics papers from the years 1963 to 2017 that explicitly deal in one or the other way with mechanisms. They describe mechanisms of the

- generation of turbulences within fluids (Bake et al. 2002);
- tsunami generation by submarine mass flows (Haugen et al. 2005);
- generation of turbulences within the accretion disk of astrophysical objects such as active galactic nuclei, quasars, or black holes (Kurbatov et al. 2014);
- pulsation of stars (pulsars) (Zhevakin 1963) and giant planets (Wuchterl 1990);
- transport, propagation or diffusion of charged particles or photons (i.e., light quanta, or energy quanta of radiation) in media such as solids, fluids, or gases;
- slowing down or acceleration of charged particles;
- interactions of subatomic particles, including particle production, in particle physics;
- the Higgs mechanism of particle physics,
- the mechanism of decoherence, in condensed matter physics.

All these examples explain the generation of a phenomenon by some kind of causal process described in terms of a physical dynamics. In cases (i)–(iv), this dynamics is a classical theory of continua. In cases (v)–(ix), it is a quantum dynamics.

5.4.1 Continuous Mechanisms

Examples (i.–iii.) are applications of continuum mechanics to fluids or gases. Example (iv.) has its origin in Eddington's theory of pulsating stars (Eddington 1917, 1926). According to this theory, which was further elaborated in recent astrophysics, pulsating stars are heat engines that produce sound waves and give rise to luminosity oscillations. Hence, examples (i)–(iv) concern classical mechanisms. They are cases of a continuous "system or structure of moving parts" [...] "resembling a machine" (to quote Collins 2012) in generating turbulences, tsunamis, or oscillations according to the laws of hydro-, gas-, or thermodynamics.

If we compare them with the recent philosophical discussion, we see that neither Salmon's account of mechanical explanations nor Glennan's approach can cope with them. Only some of the respective dynamic laws describe causal processes associated with the transmission of a conserved quantity, and none of them is fundamental.

In the case of tsunami generation, the momentum (or velocity) and density of an irrotational and incompressible flow of water are conserved (Haugen et al. 2005). In the case of turbulence generation, the flowing water is incompressible as in the former example, but a phase transition from laminar to turbulence behaviour of the water adds which is calculated by a computer simulation of random velocity perturbations (Bake et al. 2002). The turbulence mechanism in the accretion disks of binary stars (Kurbatov et al. 2014) is a similar case of a phase transition. The mechanistic explanation of the phase transition to turbulence employs precessional density waves of the accretion disk, their interactions with tidal effects in the binary star system, and the generation of unstable radial modes inducing additional density and velocity gradients in the disk. The mechanism of pulsating stars according to Eddington (1917) employs the model of a thermodynamic machine that transforms heat energy into mechanical energy. Under the assumption of a constant density of stars, the corresponding dissipation of mechanical energy gives rise to sound waves (Zhevakin 1963).

Neither Salmon's nor Glennan's approach can cope with these mechanistic explanations. In comparison to Salmon's approach, they are too complex. In particular, the mechanisms of phase transitions to turbulence and thermodynamic mechanisms are far from being simply due to the transmission of a conserved quantity. They include two-level mechanisms, in which the dynamic behaviour of the respective whole is due to causal components such as precessional density waves, tidal effects, density and velocity gradients, perturbations, etc. Glennan's claim that mechanisms are complex systems producing a certain behaviour by the "interactions of a number of parts according to direct causal laws" (Glennan 1996, 52), however, also cannot cope with the above examples. In none of the cases, the behaviour of the system as a whole is spelled out in terms of a well-defined number of causal components and "direct" (or fundamental) causal laws. The reduction of a gas or a fluid to the atoms or molecules in it does not play any role, and the explanation employs perturbational effects and phenomenological laws rather than "direct", fundamental causal laws.

Hence, both approaches are not sufficiently general to cope with sophisticated mechanistic explanations based on the classical theories of physics. Nor do they fit with my own suggestion above for the sufficient conditions of mechanistic explanations (however, I did not claim that the condition is necessary). In none of the cases discussed here does a part-whole relation expressed in terms of dynamic properties and the sum rules deriving from the conservations laws they underlie play any significant role. This is due to the fact that all of them rely on phenomenological laws that however belong to a well-established physical dynamics.

5.4.2 Quantum Mechanisms

My remaining five examples concern quantum mechanisms. In the context of the present paper, it is impossible to discuss them in detail. Let me just emphasize that they come closer to Salmon's and to Glennan's accounts of mechanistic explanations.

The mechanisms of the transport, propagation or diffusion of charged particles or photons (light quanta, or energy quanta of radiation) in media such as solids, fluids, or gases concern the propagation of (subatomic) particles in the generalized sense discussed above (in Sect. 5.3.3.3) through a complex system. They are one-level, and they are associated with the transmission of one or more conserved quantities between two or more events (Salmon 1998). The slowing down of charged particles in a massive particle detector as well as in an interstellar dust cloud, or the acceleration of charged particles by electromagnetic fields in a physics laboratory as well as by the shock waves in the plasma of an astrophysical source of cosmic rays, are not much different. (These shock waves, by the way, are just another case of the classical continuous mechanisms discussed in Sect. 5.4.2.) In all these mechanisms, direct causal laws in Glennan's (1996) sense, i.e., the fundamental laws of physics come into play.

The interactions of subatomic particles, including particle production according to the laws of quantum field theory, too, may count as one-level mechanisms deriving from the laws of a quantum field theory and respecting the conservation laws for dynamic properties such as mass-energy, charge, and spin. The case of the Higgs mechanism of particle physics is a bit peculiar, but it also just concerns the coupling of quantum fields, the conservation law for mass and energy, and in addition, a kind of phase transition concerning the spontaneous symmetry breaking of the fundamental interactions of physics.

For completeness, let me just mention that the mechanism of decoherence is only employed in papers on quantum physics belonging to the research fields of condensed matter physics, low energy physics, or physical chemistry (Dubé and Stamp 2001, Stamp 2006). The relation of these investigations to the claims of Kuhlmann and Glennan (2014) concerning the mechanistic explanations of quantum mechanics needs more investigation and is beyond the scope of the present paper.

5.5 Some Conclusions

The concept of a mechanism and the corresponding account of mechanistic explanations draw on the analogy between machines and processes in nature. Therefore, it is obviously justified to generalize them in accordance with scientific and technological progress, from the traditional mechanistic explanations based on classical mechanics to recent science. These generalizations have their counterpart in a general mechanistic methodology, which is typical of the "dissecting" sciences (Schurz 2014, 35) and which aims at decomposing natural phenomena *top-down* into lower-level causal components in order to give *bottom-up* mechanistic explanations. This methodology stems from early modern science; in particular, it has its roots in Newton's analytic-synthetic method; and it is still at work in the *top-down* and *bottom-up* approaches of current biology and neuroscience. The method of dissecting the phenomena in order to explain them in terms of mechanisms became most successful in nineteenth and twentieth century science, when physics, however, overcame the restrictions of scientific explanations to classical mechanics. Hence, mechanistic explanations today for good reasons are much more general than in the age of the traditional mechanical philosophy.

If mechanistic explanations are multi-level, they make ontological reduction possible. One of their advantages is that the ontological reduction they achieve does not need the support of a full-fledged theoretical reduction. Nor do they require that the parts of a mechanism be spatial, i.e., spatially well-distinguished or localized. The sufficient conditions for a good, heuristically fruitful mechanistic explanation are substantially weaker. In the practice of physics and its follower sciences, a sufficient condition for a good mechanistic explanation is that it establishes an empirically justified dynamic part-whole relation between a complex system and dynamic parts of that system. The mechanistic explanations of physics establish dynamic part-whole relations in terms of sum rules for conserved quantities such as mass, energy, charge, etc., in accordance with conservation laws related to the symmetries of a physical dynamics. To establish such sum rules and to give mechanistic explanations based on them is also possible in the domain of quantum mechanics, without employing the mechanism of quantum decoherence (against Kuhlmann and Glennan 2014). The current well-established constituent model of matter in terms of molecules, atoms, and subatomic particles relies on them. Hence, it is a mechanistic model in the generalized sense proposed here.

It was possible to pin down a sufficient condition for good mechanistic explanations, but it seems to be impossible to find a necessary condition for mechanistic explanations proper, as the empirical success of mechanistic explanations in cognitive neuroscience shows. Finally, my examples from current physics demonstrate that Glennan's (1996) or Salmon's (1984, 1998) physics-based accounts of mechanistic explanations are too specific. They cannot cope with cases of mechanisms based on classical continuum mechanics and thermodynamics. However, they fit better with examples from current particle physics.

References

Bake, S., et al. 2002. Turbulence Mechanism in Klebanoff Transition: A Quantitative Comparison of Experiment and Direct Numerical Simulation. *Journal of Fluid Mechanisms* 459: 217–243.

Beaney, M. 2015. Analysis. In *The Stanford Encyclopedia of Philosophy* (Spring 2015 Edition), ed. Edward N. Zalta. http://plato.stanford.edu/archives/spr2015/entries/analysis/.

Bechtel, W. 2008. *Mental Mechanisms. Philosophical Perspectives on Cognitive Neuroscience*. New York: Routledge.

Bechtel, W., and A. Abrahamsen. 2005. Explanation: A Mechanist Alternative. *Studies in History and Philosophy of Biological and Biomedical Sciences* 36: 421–441.

Bechtel, W., and R.C. Richardson. 2010. *Discovering Complexity: Decomposition and Localization as Strategies in Scientific Research*. 2nd ed. Princeton: Princeton University Press.

Buzzoni, M. 2019. Chapter 7: Multilevel Reality, Mechanistic Explanations, and Intertheoretic Reductions. In *Mechanistic Explanations in Physics and Beyond*, ed. Brigitte Falkenburg, and Gregor Schiemann. Cham: Springer.

Collins, W. 2012. Mechanism. In *Collins English Dictionary - Complete & Unabridged*, 10th Edition. HarperCollins Publishers. http://www.dictionary.com/browse/mechanism. [May 14, 2017]

Craver, C.F. 2007. *Explaining the Brain. Mechanisms and Mosaic Unity of Neuroscience*. Oxford: Clarendon Press.

Craver, C.F., and L. Darden. 2013. *In Search of Mechanisms: Discoveries Across the Life Sciences*. Chicago: University of Chicago Press.

Descartes, R. 1644. *Principia philosophiae*. Engl.: *The Principles of Philosophy*. Trans. J. Veitch. http://www.fulltextarchive.com/pdfs/The-Principles-of-Philosophy.pdf. [September 9, 2018].

Dowe, P. 2008. Causal Processes. In *The Stanford Encyclopedia of Philosophy* (Fall 2008 Edition), ed. Edward N. Zalta. http://plato.stanford.edu/archives/fall2008/entries/causation-process/.

Dubé, M., and P.C.E. Stamp. 2001. Mechanisms of Decoherence at Low Temperatures. *Chemical Physics* 268: 257–272.

Eddington, A.S. 1917. The Pulsation Theory of Cepheid Variables. *The Observatory* 40: 290–293.

Falkenburg, B. 2007. *Particle Metaphysics. A Critical Account of Subatomic Physics*. Heidelberg: Springer.

———. 2015. How Do Quasi-Particles Exist? In *Why More Is Different. Philosophical Issues in Condensed Matter Physics and Complex Systems*, ed. B. Falkenburg and M. Morrison, 227–250. Heidelberg: Springer.

Glennan, S. 1996. Mechanisms and the Nature of Causation. *Erkenntnis 44*: 49–71.

———. 2002. Rethinking Mechanistic Explanation. *Philosophy of Science 69*: 342–353.

———. 2017. *The New Mechanical Philosophy*. Oxford: Oxford University Press.

Haugen, K.B., et al. 2005. Fundamental Mechanisms for Tsunami Generation by Submarine Mass Flows in Idealised Geometries. *Marine and Petroleum Geology* 22 (1–2): 209–217.

Hobbes, T. 1655. *De corpore*. Engl. In J. C. Gaskin (ed.), *Thomas Hobbes: The Elements of Law, Natural and Politic; Ch. IV.* Oxford University Press (1999), 193–212.

Kuhlmann, M., and S. Glennan. 2014. On the relation between quantum mechanical and neo-mechanistic ontologies and explanatory strategies. *European Journal of Philosophy of Science 4*: 337–359.

Kurbatov, E.P., et al. 2014. On the Possible Turbulence Mechanism in Accretion Disks in Non-magnetic Binary Stars. *Phys. Usp.* 57(8): arXiv:1409.8492 [astro-ph.SR].

Machamer, P., L. Darden, and C.F. Craver. 2000. Thinking About Mechanisms. *Philosophy of Science* 67: 1–25.

Malisoff, W.M. 1940. Physics: The Decline of Mechanism. *Philosophy of Science* 7: 400–414.

Morgan, M., and M. Morrison, eds. 1999. *Models as Mediators*. Cambridge: Cambridge University Press.

Newton, I. [1713] 1999. *The Principia. Mathematical Principles of Natural Philosophy.* A new translation by I. Bernhard Cohen and Anne Whitman. Berkeley: University of California Press.

———. [1730] 1979. *Opticks or Treatise of the Reflections, Refractions, Inflections & Colours of Light.* Based on the fourth edition: London 1730. New York: Dover, 1952, 1979.

Pappus of Alexandria [1589] 1986. Pappi Alexandrini *Mathematicae Collectiones*. Venedig 1589. Engl. Transl.: *Book 7 of the Collection.* Part 1. Edited with Translation and Commentary by Alexander Jones. New York, 1986, 82.

Popa, T. 2018. Mechanisms. Ancient Sources. In *The Routledge Handbook of Mechanisms and Mechanical Philosophy*, ed. S. Glennan and P. Illari, 13–25. London: Routledge.

Salmon, W. 1984. *Scientific Explanation and the Causal Structure of the World.* Princeton: Princeton University Press.

———. 1998. *Causality and Explanation.* Oxford: Oxford University Press.

Schiemann, G. 2019. Chapter 3: Old and New Mechanistic Ontologies. In *Mechanistic Explanations in Physics and Beyond*, ed. B. Falkenburg and G.Schiemann. Cham: Springer.

Schurz, G. 2014. *Philosophy of Science. A Unified Approach.* New York/ Abingdon: Routledge.

Stamp, H.C.E. 2006. The Decoherence Puzzle. *Studies in History and Philosophy of Modern Physics* 37: 467–497.

Vesalius, A. 1543. *De humani corporis fabrica libri septem.* Basel (Johannes Oporinus).

Wuchterl, G. 1990. Hydrodynamics of Giant Planet Formation I. Overviewing the κ Mechanism. *Astronomy and Astrophysics* 238: 83–94.

Zhevakin, S.A. 1963. Physical Basis of the Pulsation Theory of Variable Stars. *Annual Review of Astronomy and Astrophysics* 1: 367–400.

Part II
Mechanisms, Causality, and Multilevel Systems

Chapter 6
Mechanist Explanation: An Extension and Defence

Michel Ghins

Abstract The present paper critically examines the main claims of the new mechanist account of explanation defended by Glennan and Machamer, among others. The two major difficulties of the new mechanism, namely the *circularity objection* and the *bottoming out problem* are discussed and addressed. To solve the bottoming out problem, this paper proposes a modification and enlargement of the mechanist account of explanation advocated by Salmon and Dowe. Such extension can then successfully apply to the explanations provided for the global behaviour of complex systems, not only in physics but also in other disciplines such as biology and the social sciences. The proposed enlarged version of Salmon's and Dowe's account provides a monistic account of explanation based on causal generalizations or laws which need not be construed as necessary to fulfil their explanatory role. Such change-relating generalizations or laws are characterized by the presence of a term that refers to the variation in time of some property, which is identified to the effect whereas the other terms refer to the causes. Thus, by enlarging the Salmon and Dowe notion of mechanism, I advocate a nomological theory of causation instead of the mechanistic theory of causation proposed by Glennan *et alii*.

Mechanist explanation has enjoyed a renewal of interest in the past two decades. This is mostly due to the papers by Glennan (1996, 2002), Machamer et al. (2000) and Kuhlmann and Glennan (2014) that advocate the merits of what is called the "new mechanism" or the "new mechanical philosophy" in the philosophy of science. The present paper critically evaluates the main tenets of this new mechanism. It also proposes an amendment and extension of the causal mechanism developed by Salmon (1984, 1998) and Dowe (1992, 2000) in the light of the main insights of the new mechanism. It will be argued that by relying on the notion of causal law

M. Ghins (✉)
Centre de Philosophie des Sciences et Société (CEFISES), Institut Supérieur de Philosophie, Université Catholique de Louvain, Louvain-la-Neuve, Belgium
e-mail: michel.ghins@uclouvain.be

B. Falkenburg, G. Schiemann (eds.), *Mechanistic Explanations in Physics and Beyond*, European Studies in Philosophy of Science 11,
https://doi.org/10.1007/978-3-030-10707-9_6

such extension permits to counter two major objections to the new mechanism, namely the charge of circularity and the bottoming out problem.

6.1 The New Mechanism

The main motivation for developing the new mechanist view about explanation sprang from the dissatisfaction with the accounts of explanation exclusively inspired by the explanatory practices in physics. These accounts, such as Salmon's and Dowe's, do not capture the nature of the explanations offered for the behaviour of the complex systems typically encountered in the biological and social sciences. Moreover, the new mechanists attempt to provide an explanation of the perplexing notion of causality by relying on the notion of mechanism. Stuart Glennan plainly states that he intends his "analysis of mechanisms to explain the nature of causality." (Glennan 1996, 56)

What is a mechanism? According to the new mechanism, a mechanism is a system of parts which interact in accordance with generalizations or laws. These interactions are supposed to explain the way the system behaves:

Definition 1 "A mechanism underlying a behavior is a complex system which produces that behavior by the interaction of a number of parts according to direct causal laws." (Glennan 1996, 52)

Glennan immediately remarks that a mechanism is a "mechanism underlying a behaviour":

> "One cannot even identify a mechanism without saying what it is that the mechanism does. The boundaries of the system, the division of it into parts, and the relevant modes of interactions between these parts depend upon the behavior we seek to explain. (...) A complex system has many mechanisms underlying its different behaviors" (Glennan 1996, 52)

The selection of the relevant parts and the relevant interactions in the underlying mechanism is guided by the aim to explain a specific behaviour. Different behaviours of the same complex system will be accounted for by different mechanisms. Later, Glennan modified this definition in the following way:

Definition 2 "A mechanism for a behavior is a complex system that produces that behavior by the interaction of a number of parts, where the interactions between parts can be characterized by direct, invariant, change-relating generalizations. » (Glennan 2002, S344)

Significantly, this new definition does not include the word "law" anymore, let alone the expression "causal law". The reason is that Glennan wishes to better "capture the kind of localized causal generalizations that characterize interactions between parts of mechanisms".[1] Indeed, the term "law" often has stronger and wider

[1] Private communication.

connotations, such as universality and necessity, whereas the mechanist philosophy advocated by Glennan is neutral with respect to various conceptions of law. Generalizations in his sense only describe interactions between parts, such as scissors cutting paper, wine staining cotton etc. (Glennan 2016, 810). Yet, these generalizations must "capture the relevant counterfactual truth claims" (Glennan 2002, S344), that is, the counterfactuals which are involved in our understanding of an interaction. Later, Glennan abandoned the expression "change-relating" generalisations because he doesn't want to seem committed to Woodward's views on causality (2017).

The behaviour of a mechanism occurs in time. The complex system has properties that change. If we focus on a single changing property of the whole system, the behaviour (or the phenomenon (Glennan 2016, 804)) to be explained is the temporal variation of that property by means of an underlying mechanism. What then are the component parts of a mechanism?

> "These parts must be objects, in the most general sense. They must have a relatively high degree of robustness or stability; that is, in the absence of interventions, their properties must remain relatively stable. Generally, these parts can be spatially localized." (Glennan 2002, S344)

Thus, the component parts of a mechanism are *objects*. Objects have properties which remain "relatively stable". There are two ways to understand stability here. Either the *kinds* of properties possessed by the object remain the same, such as volume and pressure for a gas, while allowing for changes of the values of these properties. Or, the *values* of some of the properties of the object remain constant. In what follows, stability will be understood in the first – more permissive – sense.

Classical mechanisms are composed of *spatially* distinct parts. Glennan does not see this as a compulsory requirement, since parts are "generally" spatially localized but they could not be. Here, he has in mind fields, such as the electromagnetic field, which cannot be confined to a small area, but are parts of many mechanisms (Glennan 1996, 56). He also keeps an eye on the applicability of his views to quantum mechanical systems whose parts are often not spatially localized (Kuhlmann and Glennan 2014). Fields then can be parts of a mechanism, and therefore objects, although they do not count as objects in the habitual sense. Particles, on the other hand, are usually considered to be objects which can interact with a surrounding field.

The interactions between parts, *i. e.* the inner workings of the mechanism, "produce" or cause the behaviour of the system, that is, the change of the properties of the system in the course of time. This is an ontological claim: the parts of the mechanism are real and the invariant generalizations are supposed to be true descriptions of real interactions. Mechanisms are the "truth makers for causal claims" (Glennan 2016, 808). Moreover, the actual behaviour of the system – the phenomenon of interest – is *causally explained* by the underlying mechanism, *i.e.* by the interactions between the component objects which produce or cause the changes of the global system's properties.

If what causes a specific behaviour of the complex system is the working of a specific mechanism, the latter is ontologically prior. The causal process undergone by the complex system is grounded on the interactions between the objects which make up the mechanism. These interactions are described by generalizations which can be epistemologically *identified* and *ascertained* without resorting to the notion of causality, in a Humean and Russellian way (Russell 1913). Besides, from an ontological perspective, the regularities described by these generalizations do not alone found the global behaviour of the system since its object-parts are also needed. Some (but not all) of the relevant generalizations are called "laws" by scientists of various disciplines (physics, biology, economics etc.) but Glennan does not want to call them "laws". Nomicity is often associated with causality, or even ground it. But, according to the new mechanists' philosophical agenda, causality should not to be based on laws, but on mechanisms, since they are rather sceptical about the notion of law or at least belittle its importance. Against them, I sustain that nomicity grounds causality (see § 6 below).

From an epistemological viewpoint, when we know the workings of the underlying mechanism we also know why the complex system exhibits a specific behaviour. A behaviour is the transition of the system from one event or state – which is defined as the occurrence of a property (or properties) – to another state. If it can be said that a preceding state of the system is the cause of a subsequent state, it is because the underlying mechanism makes the system evolve in such a way. Thus, mechanisms provide not only the ontological foundation of causal relations but causal explanation also epistemologically boils down to mechanical explanation.

To clarify the issues at stake, it is useful to distinguish three levels of causation. The first level is the *causal succession* of states of the whole system in a specific behaviour (among possible others), which is a causal process. The second level of causation consists in the workings of the underlying mechanism, namely the *causal interactions* of the relevant objects. This distinction corresponds to what Glennan calls "an important duality in mechanism talk" (Glennan 2016, 801):

> "On the one hand, mechanisms can be characterized as processes with various stages connecting initial and termination conditions. On the other hand, mechanisms can be characterized as systems – stable collections of parts or components organized in such a way as to have persistent states and dispositions."

At first sight, there seems to be a third level of causation, namely the *causation of the behaviour* of the complex system by the workings of the underlying mechanism, which causes an effect, *i.e.* the specific behaviour or phenomenon we are interested in. We would then have an inter-level causal relation between the first level (effect) and the second underlying level (cause). Let us try to make all this clearer by looking at some examples of mechanist explanation.

A first example discussed by Glennan (1996) is the float valve. The specific behaviour of the system to be explained is the maintenance (and variation) of the level of water in a tank. The relevant parts (objects) of the underlying mechanism are the tank, the water, the float and the lever, and the valve, which is connected to a pressurized water source. When the water level is low the float and the lever do not block the flow of water. As the water fills the tank the float goes up until the lever is in contact with the valve and stops the flow of water whose level then remains constant. From an epistemological (and ontological) point of view, three levels of causation are involved. The first level is the behaviour of the whole system, that is, more precisely, the variation of a property, namely the height of the water level. The variation of the height of the water is a causal process: a given height h_l is the cause of a height h_2 which occurs later in time.[2] At a second level of causation, we have the interactions between the relevant objects of the system, namely the underlying machinery. Finally, the workings of the mechanism causally produce the variation of the height of water, that is, the third (inter-)level of causation.

Let us briefly mention some other examples of mechanisms also discussed by Glennan: the voltage switch which is a mechanical device that explains the change and stability of output voltage in function of input voltage; the transmission of information between the members of a group connected by phones; the mechanical system of molecules underlying the behaviour of a gas in accordance with Boyle-Charles-Mariotte law; the mechanism of cell meiosis which explains Mendel's laws.

6.2 The Regress Difficulty with the Inter-level Causation

Glennan's mechanist account of causal explanation could lead to an unwelcome regress, which to my knowledge has not so far been noticed in the literature.[3] Glennan's main project is to found causality on the workings of a mechanism. If there is a third inter-level of causation, one should resort to a further mechanism in order to ground it. Let us call this further mechanism M_I, distinct from the mechanism M_G which is supposed to *causally produce* the behaviour of the system. Where is such mechanism to be found? And if it could be found, one still would have to explain *mechanistically* its causal action on the inter-level interaction. So, an unwanted regress looms.

Within Glennan's perspective, there cannot be room for a third inter-level interaction. True, in *Definition 2*, Glennan says that "a mechanism for a behaviour is a complex system that *produces* [our underlining] that behaviour by the interaction of a number of parts". Now, the behaviour at stake must be the behaviour of *something*. This "something" can only be the complex system itself. There cannot be two complex systems involved. So, the "something" is the complex system whose specific behaviour is to be explained. Thus, the complex system can only be the mechanism itself, that is, the system's decomposition in parts effectuated in function of the specific behaviour to be explained. In other words, the behaviour of the complex system is the behaviour of the underlying mechanism. If this interpretation is correct, there is no inter-level causation (third level), but only two levels of causation,

[2] See my definition of a causal process inspired by the Dowe-Salmon mechanicism in §6 below.

[3] I'm grateful to Brigitte Falkenburg to have pointed this to me.

namely, to repeat, the variation of some properties of the whole system (first level) and the interactions among the objects composing the underlying mechanism (second level). Such interpretation is supported by Glennan's assertion that "a mechanism is a complex system" (Glennan 2002, S344). The evolution in time of the mechanism is identical to the specific causal process undergone by the complex system with respect to the variation of the specific properties that are considered. This "definitional"[4] move permits to avert the danger of a threatening regress.

Have another look at the float valve. The complex thing is the system of the following interacting objects: the tank, the water, the float, the lever and the valve connected to a pressurized water source. The phenomenon we are interested in is the variation of one property of the global system, namely the variation of the water height in the tank. Notice that the water here is *part* of the mechanism. The behaviour we are looking at is the evolution of a property of a part of the mechanism. There is no problem in this case since interactions among parts occur at the second level of causation. The variation of the height of water is then the result of the causal interactions of the component objects of the mechanism.

Think now of the classical example of mechanism cherished by modern times mechanist thinkers such as Descartes, Locke and so on, namely a pendulum clock. The behaviour we are focusing on is the motion of the needle on a scale (which marks the time), that is, the behaviour of an object-part of the whole clock-system which is clearly causally produced by the internal workings of the clock's mechanism.

Surely, this way of apprehending the behaviour of a whole by considering only the behaviour of one of its parts cannot be applied to all complex systems. If you study the transmission of information among a group of people connected by cell phones, the behaviour of the system is captured by the variation of a property, namely the amount of information globally possessed by the system, that is, the sum of the bits of information possessed by the individuals who are connected, *i. e.* interact with their cell phones. The variation of the total amount of information is a first-level causal process. But at the second level, as the mechanism operates, the global amount of information automatically increases (or can also decrease…). It makes no sense of talking of some relation of causality between the workings of the mechanism and the behaviour of the system, just because the description of the inner workings of the system is just another way to describe the phenomenon we are interested in. This is what Glennan has in mind when he introduces his notion of a "mechanical model":

> "(MM) A mechanical model is a description of a mechanism, including (i) a description of the mechanism's behavior (ii) a description of the mechanism which accounts for that behavior" (Glennan 2002, S347)

Thus, the descriptions (i) and (ii) correspond to the first level and the second level of causation, respectively.

[4] I owe this suggestion of terminology to Peter Verdee.

6.3 The Objection of Circularity and the Bottoming Out Problem

The mechanist account of explanation faces two well-known difficulties which are openly recognized – and addressed – by Glennan: the objection of circularity and the bottoming out problem.

Let us start with the *objection of circularity*. The causal explanation of the behaviour of a complex system relies on the interactions between some of its parts, that is, the relevant ones. The notion of interaction is a causal concept. This is granted by all. Therefore, the mechanist account of causality relies on the causal concept of interaction, and we are confronted with a circularity hurdle. However, the explanation of the behaviour does not have to explicitly refer to a further (mechanical) explanation of the relevant interactions. The search for an explanation concentrates in the first place on some behaviour of the whole system, not on an explanation of the inner interactions within the mechanism.

> "In describing the mechanism that connects the two events [of the behaviour] I have explained how these events are causally connected. How the parts are connected is a different question. In can try to answer this second question by offering another account of the mechanisms that connect them, but I need not give such an account to explain the connection between the events. Indeed, such an account would only obscure the relevant features of the original explanation." (Glennan 1996, 68)

As Glennan observes, spelling out the interactions would *obscure* the mechanist explanation of the sequence of states (events) of the complex system, by making the explanation unnecessarily complicated and thus hampering understanding. So, from the epistemological standpoint, the objection of circularity is easily defeated. Explanations must stop at some point, and there is no risk of circularity or regress here.

The situation is more complicated with the *bottoming out problem*. One could indeed ask how the interactions between the parts of a mechanism work and attempt a mechanical explanation of these interactions, such as the interaction between the water and the float in the above example. To achieve this, we would have to divide some parts of the system into subparts and explain the generalizations ("laws") describing the higher-level interactions in terms of lower-level interactions between these subparts. Going further down, we would eventually reach the microphysical level where interactions are governed by the fundamental laws of physics for which no further mechanist explanation is available, at least at the present stage of scientific knowledge. This is the case for electromagnetic interactions since the electric and magnetic fields cannot be divided into independently manipulable parts. The space-time points, at which there is a value of the electric and magnetic fields, are not objects because objects must be capable of interacting and they also have to be manipulable constituents (Glennan 1996, 56).

True, explanation must stop at some point, and there is no threat of an unwelcome regress or circularity there. But as far as the ontological basis of causality is concerned, there are reasons to worry. To causally explain the regularity of a behaviour, which can be described by a universal statement or lawlike sentence, is to appeal to a real mechanism which grounds such regularity. By definition, if a generalization or a law is fundamental, there is no underlying mechanism available, at least for now. At any rate, even if the laws considered to be fundamental at this time turn out to be based on mechanisms, in practice we will always hit a last ground in the hierarchy of groundings and thus a bottom of fundamental laws, which might be provisional or final, but is nevertheless a bottom. Therefore, either fundamental laws are not causal or, if they are, they are not mechanistically founded. In either case, the new mechanism is in trouble. In the first situation, some regularities cannot be causally grounded. In the second situation, we would have cases of causation which are not mechanist and this would contradict the very aim the mechanist approach.

Glennan acknowledges that there is no mechanical explanation of fundamental laws (otherwise, they would not be fundamental…). Thus, at the fundamental level there is no causal explanation of regularities and the problem of distinguishing generalizations which are causal - and would deserve to be called "laws" (again, in the sense of generalizations which ground the truth of relevant counterfactuals) - and which generalizations are not nomological remains acute. Glennan offers no satisfactory solution to this problem. From an epistemological point of view, this is not a serious difficulty, since explanations must end at some point, and we are not required to explicitly formulate the generalizations which describe the interactions between parts to causally explain a specific behaviour of a complex system. Ontologically, however, causation at higher levels is not ultimately grounded on causation at the fundamental level, since the mechanist account of causation is simply unavailable at the level of the fundamental laws of physics. Glennan then is forced to accept that the mechanist construal of regularities is *ontologically dualistic*. On the one hand, we have causal regularities grounded on actual mechanisms; on the other hand, the most basic regularities described by the fundamental laws of physics must be considered to be causal as well, but in a distinct, non-mechanist sense, on pains of becoming incoherent with respect to the mechanist account of causality.

> "This explanation of the role of mechanisms is available as long as the generalizations are mechanically explicable, but here we come to what may seem the key *metaphysical*[5] issue. If mechanisms are truly going to explain how one event produces another, all interactions between parts, at all levels of the hierarchy of mechanisms, will need to be genuinely causally productive. If it were to turn out that these interactions at the fundamental level were not truly interactions, then none of the putative causal interactions mediated by mechanisms would be genuine." (Glennan 2011, 811)

Glennan needs causality all the way down, simply because causal interactions at all levels are real. Ontologically or metaphysically, causal relationships at one level must be based on real causal relationships at the lower level. His new mechanism is foundationalist in this sense.[6] If Glennan adopts such ontological posture, this is not because he believes that causation is grounded on some kind of necessity. It is

[5] Our underlining.

[6] On this we disagree with Laura Felline who claims that Glennan's "mechanist account of causation might be consistent with a no-causality-at-the-fundamental level view." (2016, 251–262) Glennan's new mechanism is ontologically foundationalist.

sufficient for him that the fundamental laws (conceived as "invariant change-relating generalizations") warrant the reliability of the relevant counterfactual conditionals. He does not elaborate however on the connection between laws and counterfactuals, but it seems that according to him such connection would be sufficient to have causal interactions at the fundamental level. Be it as it may, such construal of causality at the fundamental level would certainly not be mechanical in his sense. This is why Glennan grants that his view on causality is *dualistic*.

Is it possible to avoid this dichotomy between mechanist causality and fundamental causality? I will try to show that a unifying, monistic account of causality can be defended provided we enlarge the causal-mechanist view advocated by Salmon and Dowe, which mostly is applicable to physics.

6.4 The Salmon-Dowe Mechanism

As Peter Railton claimed back in 1978:

> "The goal of understanding the world is a theoretical goal, and if the world is a machine – a vast arrangement of nomic connections – then our theory ought to give us some insight into the structure and workings of the mechanism." (Railton 1978, 208; quoted by Salmon 1984, 123)

Wesley Salmon makes Railton's insight more explicit:

> "Causal laws *govern* the causal processes and causal interactions, providing the regularities that characterize the evolution of causal processes and the modifications that result from causal interactions. Causal processes, causal interactions, and causal laws provide the mechanisms by which the world works; to understand *why* certain things happen, we need to see *how* they are produced by these mechanisms." (Salmon 1984, 132)

It is fair to contend that, according to Salmon, a machine or a mechanism is a system of causal processes and causal interactions at the intersection of processes. A first important difference between Glennan's and Salmon's mechanisms is the following. For Glennan, mechanisms are composed of *stable* things or objects (Glennan 2002, S345), whereas for Salmon, mechanisms are composed of sequences of events or causal processes which intersect, and needn't exhibit the kind of stability that objects have. Yet, for Salmon, processes are worldlines of stable objects, such as particles of bodies. For Salmon, the aim is to explain why certain events ("things", as he says) happen by means of the causal processes and causal interactions which constitute the parts of the mechanism. Thus, for both Glennan and Salmon, the behaviour to be explained would be a sequence of events, that is the temporal occurrences of properties along a space-time line, or the occurrence of a particular event, such as the breaking of a window or the death of the Archduke in Sarajevo. As Glennan says, we attempt to explain the mechanism as a process, namely its trajectory from an initial to a final state, by means of a system of interactions between its inner parts.

However, Salmon does not pay much attention to the global states of some mechanical systems as wholes. Here, we have a second main difference between Glennan's new mechanism and the mechanism promoted by Salmon.[7] For the latter, there is only one level of causality along individual space-time lines and at their intersections. New mechanists, on the other hand, aim at providing causal explanations of what is going on at the level of complex systems (first level) on the basis of the interactions of its parts (second level). My point is that it is possible to unify both approaches by considering the new mechanist' complex systems as mechanisms of causal processes and causal interactions, which underline the behaviour or causal process of the complex system as a whole. Just because the behaviour of an object in Glennan's sense is a causal process in Salmon's view, provided we don't consider physical properties only.

Let us look again at the float valve. Glennan's main concern is to mechanistically explain the behaviour of the complex system *in toto*, whereas Salmon concentrates on the causal processes of the parts and the causal interactions between them. Yet, the two approaches can be reconciled. In this example, the behaviour of the whole system is characterized by a single property, namely the height of the water level in the tank. According to Salmon, there is a mechanism underlying this behaviour, namely the causal processes of the various objects involved and their causal interactions. More generally, the behaviour of a complex system is a temporal sequence of states, each of which is specified by a number of properties. The occurrence of a state at some time is an event. If this is so, the behaviour of a complex system is a sequence of events characterized by the temporal variation of some properties. Such behaviour is explained, both by Salmon and Glennan, by the workings of a mechanism. Glennan speaks about the interactions of the component objects according to invariant change-relating generalizations, whereas Salmon appeals to the notions of causal processes, causal interactions and causal laws. But later on, reacting to objections addressed by Phil Dowe, Salmon attempted to avoid the recourse to causal laws (Salmon 1984), just as Glennan did.

How are the crucial notions of "causal process" and "causal interaction" to be defined? Here is the definition provided by Salmon after he took into account some of Phil Dowe's (1992) critical remarks:

> "A causal process is a world-line of an object that *transmits*[8] a non-zero amount of an invariant quantity at each moment of its history (each space-time point of its trajectory)." (Salmon 1998, 257)

According to Salmon and Dowe, such definition permits to distinguish processes which are causal form processes which are not and which are called "pseudo-processes", like the worldline of the shadow of a moving car. Since transmission is a causal notion, Salmon proposes an allegedly *non*-causal definition of transmission to avoid the charge of circularity:

[7] I'm grateful to Brigitte Falkenburg for having drawn my attention to this difference.

[8] Underlining is mine.

> "A process transmits an invariant (or conserved) quantity from A to B (A≠B) if it possesses that quantity at A and at B and at every stage of the process between A and B without any interactions in the half-opened interval (A, B] that involve an exchange of that particular invariant (or conserved) quantity." (Salmon 1998, 257)

This is Russell's "at-at theory of causal transmission", reformulated by Salmon.

"A causal interaction is an intersection of world-lines which involves the exchange of a conserved quantity" (Dowe 1992, 212, and also Salmon 1998, 253).

In virtue of these definitions, an event is explained when it is shown that it is inserted in a nexus of causal processes and interactions. The component parts of a mechanism are the objects which manifest causal processes and interact when their worldlines cross together with an exchange of a conserved physical quantity. This way of explaining is called an "ontic account of explanation" in the literature since it relies on the existence of a causal network in the world. A sequence of events, which can be a regularity, is thus explained by an underlying mechanism.

The examples discussed by Salmon and Dowe are exclusively drawn from physics, such as the collision of two billiard balls and atomic decay. According to them, if you consider the system of two billiard balls, there are two causal processes along which the linear momentum (quantity of motion) is conserved. When the balls collide, an exchange of linear momentum takes place at the intersection of the worldlines of the balls (Dowe 2000, 102). No attention is paid to the properties of the global system of the balls. Glennan and the new mechanists rightly complains about this. They also criticize the narrow physicalist strategy implemented by Salmon and Dowe. But these defects can be eliminated by drawing attention to the properties (which needn't be physical) of whole systems whose internal workings are captured by mechanisms as understood by means of an extension of the Salmon and Dowe account of explanation, as will be shown below.

Before we do that, it is appropriate to point out two main difficulties encountered by Salmon's ontic account of explanation. As we observed, Salmon and Dowe strive to avoid the notion of law. However, conservation of an invariant quantity without interaction along a worldline is a nomological feature of the world described by conservation laws. Moreover, interactions are ruled by laws. To capture the nature of a particular interaction it is not enough to say that an interaction is by definition the exchange of a conserved quantity at the intersection of worldlines. One must also formulate the precise law that governs this exchange when we offer an explanation in a specific case. What is more, laws governing interactions are causal laws and the threat of circularity looms again…

A second difficulty arises from the fact that a worldline is always the worldline *of an object*, such as a billiard ball. How do we define the identity of an object through time? Averting circularity forbids any recourse to causal notions. Biting the bullet, Dowe claims that the notion of identity of an object is a "primitive concept" (Dowe 2000, 107). But, as Psillos remarks, this is "deeply unsatisfactory" because then "which processes count as causal will also be a primitive concept of his theory." (Psillos 2002, 123–124)

Before proposing a modified and enlarged version of the Salmon-Dowe causal mechanist explanation which addresses these two difficulties, it will be profitable to make a brief historical detour to early modern science and the origins of mechanical explanation.

6.5 Cartesian Intermezzo

The classical early modern science mechanism aimed at explaining phenomena, *i.e.* "sequences of observed events", by means of mechanisms. Typically, the behaviour of a clock is explained by the internal motions and interactions of wheels, weights, chains etc. which produce the regular motion of a needle on a scale. Surely, these ancient mechanists intended to mechanistically explain the behaviour of whole complex systems, just as the new mechanists do.

For the founding fathers of modern science such as Galileo Galilei, René Descartes and Robert Boyle among others, a mechanism is a spatially organized ensemble of bodies in relative motions. Bodies possess the properties of extension (volume) and geometrical shape. Besides, they are hard (impenetrable) and interact by contact or collision. Thus, a mechanism is a complex system the behaviour of which is characterized by sequence of states (events), that is, the occurrence of properties along a worldline. The relevant properties, the celebrated "primary" properties, are distinguished by two main features: they are objective and they have a quantity which is measurable in principle. The "secondary" qualities on the other hand, are deemed to be subjective and supposed to be explainable by the behaviour of corpuscles endowed with primary qualities only.

The laws of nature, which originally were limited to the laws of motion, are relations between the primary properties of bodies such as position, velocity, extension, shape etc. As history of science unfolded, the scope of primary properties, *i. e.* the scientifically relevant properties, has been considerably extended to include temperature, mass, electromagnetic field, amount of information, DNA splitting and so on. For our present purpose, it is important to stress that laws state relations between *properties*. I'm happy to grant that properties are properties of objects (or corpuscles according to the early modern science mechanists). But we do not have to enter an ontological analysis of objects to understand what a scientific law is. Within a scientific context, it is plausible to maintain, as Hume and many others do, that an object is only a bundle of properties. This is not the way I take objects to be.[9] But for a discussion of mechanical explanation we can safely concentrate on properties and leave aside an analysis of the concept of object.

---

[9] I do not defend myself that an object is a bundle of properties. An object is an entity that has properties and can stand in relations with other objects. My position is close to the one advocated by John Heil (Heil 2003, chapter 15).

Within the objectifying approach specific to science, an entity is reduced to an object, *i.e.* a bundle of properties which can sometimes be organized according to a mathematical law. Thus, a (perfect) gas is reduced to a collection of properties such as volume, pressure and temperature related in accordance with the Boyle-Charles-Mariotte law. The behaviour of a clock is reduced to the circular motion of a needle on a scale that can be described by a general statement. Properties are judged to be objective when their measurements by different individuals and apparatuses at different spatiotemporal locations deliver the same results. In scientific activity, we can rest satisfied with objectivity in the weak sense of intersubjective agreement. In science, an object is a collection of objective and measurable properties. The relations between those properties might be governed by a scientific law.[10]

Surely, the behaviour of a clock, as well as the behaviour of a gas, can be explained by the interactions between their component parts that are bundles of primary qualities, which interact only by contact.

6.6 The Salmon-Dowe Mechanism Amended and Expanded

Our historical detour to the birth of modern science taught us that mechanist explanation eventually relies on properties and laws construed as relations between properties. The concept of object is no longer primitive since an object can be considered as nothing else than a collection of properties (which needn't be physical only). Furthermore, we cannot avoid appealing to laws to accurately describe the interactions involved in particular cases. Given this, we are now in a position to propose the following definition of a mechanism.

A mechanism is a complex system of causal processes and causal interactions – described by causal laws - which explain its behaviour, i. e. the changes of some of its global properties with time.

The parts of a mechanism are objects, *i. e.* bundle of properties, which remain constant or vary along worldlines and at their intersections in accordance with causal laws. The central explanatory role evidently rests on causal laws. We must then say what is a causal law. Moreover, we must define the notions of process, causal process and causal interaction.[11]

A process is a continuous worldline (space-time line or tube) along which some properties are conserved or vary according to a process law.

The properties at stake are not just physical properties but can also be properties referred to in scientific disciplines such as biology, economics, demography etc. However, the recourse to a law is necessary to distinguish between worldlines which are processes and which are not. A process is not just any line drawn in space-time along which some properties vary. There must be a *connection* between the events in a process and such connection is achieved by a law. We thus avoid the problem of

[10] This is not necessary however: the charge and spin of the electron are not related by a law…

[11] For further discussion of these concepts, see Blondeau and Ghins (2012).

the identification of objects. But what about the identification of properties? Properties are identified in the first place in perception, which permits us to distinguish different properties and to attest that a property remains constant or vary along a worldline. A detailed discussion of the identity of properties would take us to far afield. On this issue, I follow Heil for whom the identity and similarity of properties is a primitive relation (Heil 2003, chap. 14), although I include in the list of relevant properties all scientific properties, whereas Heil doesn't.

The word "tube" occurs in the definition to allow for regions of space-time endowed with some properties. (In what follows we will use the word "worldline" to refer to tubes as well as lines). For example, the worldline of the water in a tank is characterized by the permanence of the property of being water. (It is a law that molecules keep their structure unchanged in the absence of chemical reactions.) Along this worldline, the height of the water varies in a regular fashion. Such regularity holds for *all* tanks equipped with a float valve, a water source etc. which are organized in a certain structure. A universal statement describing this regular behaviour of water (given some initial conditions) could then be considered as a (non-natural) phenomenological law related to the relevant counterfactual conditionals. At any rate, this regularity is governed by several process laws, namely the laws that describe the internal processes of the underlying mechanism.

A process law is a mathematical law which contains time as one of its variables.

The restriction to mathematical laws is not compulsory however. Process laws could be formulated in natural language by means of universal sentences that make reference to time. This is the case for some change-relating generalizations used in biological and social sciences. A conservation law is a law which asserts that some property does not vary in time.

Some processes are causal, whereas some are not. For a process to be causal, an additional condition is required: a *causal process* is a process governed by a causal law. What then is a *causal law*?

A law is causal if and only it contains a time derivative.

Such definition was originally proposed in physics by Philipp Frank in 1932 (Frank 1997). (Again, a causal generalization can be formulated in natural language. It is causal if and only if it contains words that refer to some change.) This criterion is formal and intrinsic (see Blondeau and Ghins 2012). The effect corresponds to the time derivative of a quantity (property) in the mathematical expression of the law; the other term(s) designate the cause(s). Such definition conforms our intuition: a cause is what produces a change, namely the variation of properties in time, that is, an effect.

A causal interaction is an interaction between distinct processes, which is described by a causal law and which results in changes of properties along the involved processes.[12]

[12]There is causation along a causal process, without interaction. In an inertial process for example: the variation of position is caused by the constant velocity (see Blondeau and Ghins 2012, 398). Thus, causality is not restricted to causal interactions.

Such definition does not require that interactions be local.[13] In a causal law, an instantaneous change in time is mathematically expressed by a time derivative.

$$E = \frac{\partial x}{\partial t} = C_1 + \ldots + C_n$$

E denotes the effect: the variation of the quantity x in time. C_i refer to the causes – forces, fluxes, sources etc. – the sum of which is equal to the effect. C_i do not denote variations in time, but they may be functions of time. If C_1, C_2,.... are all equal to zero, or if their sum is zero, x does not change, and we have a conservation law. Examples of causal laws include Newton's second law, the law of chemical kinetics, the Schrödinger equation, some laws of demography, the Lotka-Volterra laws of evolution of predators and prey, the Kermack-McKendrick equations in epidemiology, the Vidale-Wolf equation in marketing etc. (Blondeau and Ghins 2012). The proposed criterion of causal nomicity is not limited to physics but applies to other scientific disciplines as well. The proposed extended mechanism can be applied to mechanisms which are biological, social etc. all the more that change-relating generalizations or causal laws could be formulated in natural language in which some terms or expressions refer to effects and others to causes.[14]

In reply to Glennan's concern about the explanation of the behaviour of whole systems, this *Enlarged version of the Salmon-Dowe causal Mechanism* (hereafter ESDM) conforms to the explanations given for the global behaviour of complex systems, whether stable or not. This is clear enough for the variation of the level of a tank, where the interaction between the worldline of the water and of the flow is governed by the laws of fluids. The behaviour of a gas is classically explained by the causal processes and interactions of its component molecules. Each molecule is associated to a worldline which is a causal process governed by Newton's laws. When a collision occurs, the interaction obeys Newton's second law. For gases, their macroscopic properties are related to the average properties of the molecules by means of mathematical equations.

For more complex systems, such as biological or social systems, a complete causal mechanist explanation in terms of fundamental laws is not articulated in most cases, nor is it always feasible or even necessary. Such complete explanation is not needed to understand how the underlying mechanism explains the behaviour of the system. It would even make the explanation unnecessarily obscure or convo-

---

[13] This account leaves open the possibility of non-local interactions.

[14] Glennan objects [private communication] that our criterion of lawhood could be applied to correlations due to common causes, such as the correlation between the fall of a barometer and the occurrence of a storm. But there is no process connecting these two events, and a fortiori no causal process connecting them. The correct explanation of the storm is of course the occurrence of a depression. There is a causal process connecting the depression and the fall of the barometer, and another causal process connecting the depression and the occurrence of the storm. These causal processes are described by causal laws according to our criterion. Empirical inquiry allows us to detect genuine processes and causal processes from apparent ones. Glennan appears to agree with this kind of strategy since he believes that we can distinguish between genuine activities from other happenings by relying on empirical investigation (Glennan 2016, 810).

luted. A main virtue heralded by the proponents of the mechanist explanation is precisely that intelligibility and understanding of the production of some behaviour is achieved without the formulation of laws (Machamer et al. 2000, 4). Nevertheless, some change-relating generalizations which describe the relevant causal processes and the interactions that take place between the parts of mechanisms must be indicated, otherwise we would not understand how the mechanism functions. For example, the mechanist explanation of protein synthesis resorts to the following change-relating (causal) generalization: in appropriate triggering circumstances the DNA double helix splits. Splitting is a causal process whose end point is the separation of the two complementary sequences of bases. Besides, going further down, there is every reason to believe that laws play a crucial role at the fundamental level. In the mechanisms of chemical neurotransmission and protein synthesis examined by Machamer et al. (2000), it would be hard to deny that laws play a crucial role for an illuminating explanation of the relevant behaviour. At bottom, quantum mechanical processes certainly are involved.

6.7 The Circularity Objection and the Bottoming Out Problem

Does this expanded version of the Salmon-Dowe causal mechanist approach solve the difficulties faced by the new mechanist approach, namely the circularity objection and the bottoming out problem?

As we saw above, the circularity objection is easily defeated by Glennan, who points out that we can causally explain the behaviour of a whole system by resorting to causal relations between the component parts of the mechanism. Thus, two levels of causation are kept carefully distinct: the level of the global behaviour and the level at which the interactions between the relevant parts operate. Once we have described the inner workings of the mechanism which produce the global behaviour, the explanation comes to an end. Analogously, for our ESDM approach, the explanation for a behaviour relies on causal processes and causal interactions. The explanation is achieved when the causal processes and causal interactions, which are responsible for the behaviour, have been described. Instead of referring to objects and interactions between them, the ESDM view involves properties, causal processes and interactions between these processes.

Within the approach I propose, the bottoming out problem is solved thanks to the concept of causal law. A complete explanation of the behaviour of a mechanism ultimately rests on fundamental causal laws.[15] Causal laws are identified by an intrinsic, formal criterion, as suggested above. The effect is identified as the variation of some property in time. The behaviour of a mechanism is also characterized by such temporal variation of properties. The cause of such behaviour is to be found

[15] Even in explanations of an emergent behaviour, we cannot do without causal laws.

in the inner workings of the mechanism, which according to ESDM boil down to causal processes and causal interactions described by causal laws. If this approach is correct, our enlarged approach provides a unified, monistic account of causality, unlike Glennan's dualistic view.

At the fundamental level, the concept of causality is captured by causal laws, as defined by the proposed intrinsic criterion. At higher levels, the workings of mechanism are also described by causal laws, *i. e.* change-relating generalizations, which describe the behaviour of mechanisms. It must be stressed that no metaphysics of laws is needed to articulate the ESDM conception of explanation. Laws can only be considered to be general statements which describe regularities related to counterfactual conditionals. In such explanatory context, we can refrain from attributing any kind of necessity to laws. The explanatory role of laws does not rest on any kind of necessity.[16]

6.8 Conclusion

The new mechanism has two main merits. First, it has drawn our attention on the explanation of the global behaviour of complex systems. Second, according to the new mechanism, mechanical explanation is not limited to simple physical systems and interactions but can in principle be applied to the complex systems investigated in all scientific disciplines. We have attempted to show that the Salmon-Dowe causal mechanist account of explanation can be amended and extended in such a way to generally comprehend the nature of the causal explanations offered in practice for the global behaviour of complex systems in all scientific fields. Moreover, we have proposed a unifying solution of the so-called "bottoming out problem" by resorting to the notion of causal law, understood as a universal statement which mentions the variation in time of some properties of the system in function of other properties. For the purpose of explanation, those laws can be construed as mere "invariant change-relating generalizations" which do not involve any kind of necessity.

Acknowledgements I wish to express my gratitude to Mario Alai, Gregor Schiemann and especially Brigitte Falkenburg and Stuart Glennan for their careful reading and valuable suggestions, which led (I hope) to an improved version of this paper. I also thank the participants in the "Work in progress seminar" (WIP) of the "Centre de Philosophie des Sciences et Sociétě" (CEFISES) especially Alexandre Guay and Peter Verdee for their stimulating feedback.

[16]This position does not prevent to defend the necessity of laws on the basis of some metaphysics for reasons which are distinct from their explanatory capacity. In my view, only necessary laws can ground the truth of counterfactual conditionals since the latter involve modalities (see Ghins 2007, 2014)

References

Blondeau, J., and M. Ghins. 2012. Is There an Intrinsic Criterion for Causal Lawlike Statements? *International Studies in the Philosophy of Science 26*: 381–401.

Dowe, P. 1992. Wesley Salmon's Process Theory of Causality and the Conserved Quantity Theory. *Philosophy of Science 59*: 195–216.

———. 2000. *Physical Causation*. Cambridge: Cambridge University Press.

Felline, L. 2016. Mechanist Causality and the Bottoming-out Problem. In *New Directions in Logic and the Philosophy of Science,* ed. Felline, L., Ledda, A., Paoli, F., Rossanese, E. Proceedings of the Triennial International Conference of the Italian Society for Logic and the Philosophy of Science (SILFS), 18-20 June 2014, University "Roma TRE", 257–266. London: College Publications

Frank, Ph. (1997), *The Law of Causality and Its Limits*, ed. by R. S. Cohen. Berlin: Springer

Ghins, M. 2007. Laws of Nature: Do We Need a Metaphysics? Fifth Principia International Symposium. *Principia 11*, 127–149. http://www.cfh.ufsc.br/%7Eprincipi/p112-3.pdf

———. 2014. Causal Powers as Metaphysical Grounds for Laws of Nature. In *Proceedings of the AIPS Conference in Široki Brijeg (Bosnia-Herzegovina), 24–27 July 2013. Epistemologia*, ed. E. Agazzi, 83–98. FrancoAngeli: Milano.

Glennan, S. 1996. Mechanisms and the Nature of Causation. *Erkenntnis 44*: 49–71.

———. 2002. Rethinking Mechanistic Explanation. *Philosophy of Science 69*: 342–353.

———. 2011. Singular and General Causal Relations: A Mechanist Perspective. In *Causality in the Sciences*, ed. P.M. Illari, F. Russo, and J. Williamson, 789–817. Oxford: Oxford University Press.

———. 2016. Mechanisms and Mechanical Philosophy. In *The Oxford Handbook of Philosophy of Science*, ed. P. Humphreys, 796–816. Oxford: Oxford University Press.

———. 2017. *The New Mechanical Philosophy*. Oxford: Oxford University Press.

Heil, J. 2003. *From an Ontological Point of View*. Oxford: Oxford University Press.

Kuhlmann, M., and S. Glennan. 2014. On the Relation Between Quantum Mechanical and Neo-mechanist Ontologies and Explanatory Strategies. *European Journal of Philosophy of Science 4*: 337–359.

Machamer, P., L. Darden, and C.F. Craver. 2000. Thinking About Mechanisms. *Philosophy of Science 67*: 1–25.

Psillos, S. 2002. *Causation and Explanation*. Montreal: McGill-Queen's University Press.

Railton, P. 1978. A Deductive-Nomological Model of Probabilistic Explanation. *Philosophy of Science 45*: 206–226.

Russell, B. 1913. On the Notion of Cause. *Proceedings of the Aristotelian Society 13*: 1–26.

Salmon, W. 1984. *Scientific Explanation and the Causal Structure of the World*. Princeton: Princeton University Press.

———. 1998. *Causality and Explanation*. Oxford: Oxford University Press.

Chapter 7
Multilevel Reality, Mechanistic Explanations, and Intertheoretic Reductions

Marco Buzzoni

Abstract In the first part of this paper, I argue that what we consider to be a "mechanism," "level," or "component" depends on the perspectives that scientists have explicitly or implicitly adopted. If this is right, interlevel explanations and intertheoretic reductions become intimately connected. That is, I show that we cannot make sense of competing interlevel explanations of the same phenomena without reference to how those different levels of analysis relate or reduce to one another. And likewise, we cannot make sense of intertheoretic reductions (e.g., Newtonian to Relativistic equations of motion) without reference to how explanations of the same phenomena produced by those different perspectives compare. To better understand this connection between interlevel explanations and intertheoretic reductions a distinction is drawn in the second half of this paper, though only provisionally, between strong and weak relations. In the end, this distinction is removed as it only suggests ideals which are never reached in practice. Nevertheless the distinction helps us to understand how scientists can relate multiple perspectives to one another and also why they should always ultimately try to seek out newer, wider and deeper perspectives. In the end, I argue that how to relate interlevel explanations and complete theoretical reductions in science cannot be answered in the abstract by purely philosophical considerations, but only with reference to, and in accordance with, the actual history and practice of science. This is true not only for physical, but also for all empirical theories, something I show with a brief example from current cancer research.

M. Buzzoni (✉)
Department of Humanistic Studies, University of Macerata, Macerata, Italy
e-mail: marco.buzzoni@unimc.it

B. Falkenburg, G. Schiemann (eds.), *Mechanistic Explanations in Physics and Beyond*, European Studies in Philosophy of Science 11,
https://doi.org/10.1007/978-3-030-10707-9_7

7.1 Introduction

Although recent mechanistic approaches take up and develop the tradition of classical mechanical philosophy, they decidedly reject both the deductive-nomological model of explanation and a reductive view of science. Descriptions of mechanisms, in which mechanistic explanations essentially consist, have to be as multilevel as the phenomena they are invoked to explain (cf. for example Machamer et al. 2000: 13; Craver 2001: 63; Glennan: 2002; Bechtel 2008: 146–147).

In this connection, the possibility of interlevel explanation (and causation) is a crucial question for the new mechanistic philosophy, and this, in turn, requires, among other things, a satisfactory definition of the "sameness of level." In Sect. 7.2, I shall discuss one objection that has been raised by Eronen 2013 and 2015 against Craver's account of what it is to be at the same level. As we shall see, the objection is unsound though worth examining, since the correct answer to the objection is important for developing coherent notions of "mechanism," "level," and "component."

I shall endeavour to show that, if these notions are influenced by the pragmatic interests and practical possibilities of the scientists, they can be interpreted in a perspectivalist way, which is not open to the objection in question. Language can never produce a copy of reality because – as should be obvious after Weber, though it is often forgotten – reality can only be explored from particular perspectives which depend upon different pragmatic interests.

Here the key notion is that of "perspective", which may be provisionally defined as the relationship in which concrete human beings stand to their environment and which always partakes at once (1) of our acting and interacting in (and with) the world, and (2) of the inevitable theoretical mediation, both of this acting and of its repercussions in the world – for the sake of simplicity, we confine ourselves to (human) empirical knowledge and exclude philosophical discourse (for which see Buzzoni forthcoming).

Some aspects of this definition, which themselves would need definition, will be indirectly clarified in what follows, although it cannot be the aim of this paper to give a comprehensive outline of "perspective".[1] However, before proceeding further, it will be well to mention that our use of the term "perspective" applies both to simplest observation and the most powerful theories. A table can be examined as a mechanical phenomenon as well as a chemical one. In particular, a mechanical phenomenon results from considering the table from a partial point of view, which takes into account only some properties of it, such as force, mass and certain spatial and temporal relations. In turn, by further specifying the perspective under which we consider a mechanical phenomenon, for example by abandoning the principle of invariance of mass, two instances of what was considered the 'same' physical

[1] For more details see Buzzoni 2016, in which in turn I draw on the "thesis of the perspectival character of scientific knowledge" to be found in Buzzoni 1995 (*passim*, especially pp. 209–210) and 1997 (passim, especially pp. 25, 30, and 40).

phenomenon can become two different phenomena. In a similar sense, the same mechanism can become the subject-matter not only of multiple autonomous disciplinary perspectives, but also of different theories from within the same discipline. Because different disciplines and different theories of the same discipline approach problems from different perspectives using different concepts and techniques, they bring to light different mechanisms, or different levels or components of mechanisms, which are, prima facie and in a sense still to be examined, independent of one another.

Many other scholars have taken a similar path by stressing the context-relativity of mechanisms and levels and by putting them in relation to pragmatic factors, such as different background knowledge and/or different knowledge interests. However, a perspectival conception according to which different explanatory levels correspond to different perspectives is only a necessary, not a sufficient condition of a satisfactory account of interlevel explanation, since such a conception cannot yet explain the interactions and clashes between different interlevel explanations that attempt to solve the same problem(s). If each mechanism, level or component is relative to a completely distinct perspective or mode of conceptualization, it is difficult to see how there can be room both for causal interactions between entities of two different ontological levels of a certain mechanism and for alternative explanations that are provided in terms of entities belonging to different and autonomous levels. Or, to put the same point in a different way, if different theoretical perspectives pick out different mechanisms, levels, and components, an acceptable solution to the question of the *interlevel* interactions and explanations goes hand in hand with that of the old and difficult problem concerning the nature of *intertheoretic* reduction (and *vice versa*).

A simple example, which will be discussed in Sect. 7.4 more in detail, may help to illustrate the connection between the two problems. Protein folding research passed through different phases. During the first phase, research was essentially carried out from the perspective of in silico and in vitro studies, while the second phase in protein folding research was characterised by in vivo studies concerning the influence of other biological molecules in the cellular environment. Now, in vivo studies are much more complex than in silico or in vitro ones: they are a new perspective, which is wider than the simple application of knowledge already gained by in silico and in vitro methods. But this broader perspective requires to be related to all of the physical and chemical considerations concerning protein folding that still apply to some degree to the study of aggregating proteins, and this relating is undeniably an important feature in common with *intertheoretic* reduction relations.

The problem concerning the nature of intertheoretic reduction is addressed in the second part of the paper. More precisely, in Sects. 7.3, 7.4, and 7.5 I approach the problem by distinguishing, provisionally and ideal-typically,[2] two sorts of relations

[2] I use the expression "ideal-typically" in one of Weber's senses, according to which the correctness of ideal types is not necessary for their use because they are literally false but still useful models of the reality we are striving to understand. More precisely, ideal types may function "negatively" because of their heuristic significance for the assessment of reality. For example, the

between knowledge claims (be they interlevel or intertheoretic): a weak and a strong one. The first holds among prima facie different but incommensurable (and therefore compatible, since validated by different, but equally possible assumptions or criteria) theories, and the second among theories which are in competition because they provide alternative explanations of the same phenomena. It will turn out that, contrary to what was assumed by the initial definition, 'weak' (interlevel or intertheoretic) relations are not really as weak as they may initially appear, and 'strong' reduction relations are not really as strong as they may initially appear. As a result, this distinction, at least in a sense, must be abandoned. The two kinds of reduction (strong and weak) have in common that relating multiple perspectives to one another in order to better understand the subject-matter under investigation requires constructing a new, wider or deeper perspective. Furthermore, the construction of new, wider or deeper perspectives will always take place in a specific historical and practical context to suit the needs of science, and cannot therefore be the work of philosophers (Sect. 7.6). Finally, I shall show, however cursorily, that this is arguably true not only for physical, but also for all empirical theories, by presenting an example from current cancer research (Sect. 7.7).

The most obvious methodological moral is that while researchers must always be selective and prepared to use several, multilevel perspectival accounts to inform one another in connection with specific problems (cf. O'Malley et al 2014, who also comes to this conclusion), they must also be integrative, in the sense of trying to establish some interrelations – sometimes weaker, sometimes stronger – holding between different theoretical perspectives and, at the same time, between the ontological mechanisms, levels and components postulated by those perspectives.

7.2 Perspectivalism, the New Mechanistic Philosophy, and the Problem of Interlevel Explanations

Apart from some other differences that are not directly relevant in the present context, recent mechanistic approaches essentially agree in conceiving a mechanism as defined by Machamer et al. 2000, that is, as a device consisting of interrelated parts that, starting from an initial situation and ending with a final result, performs some kind of work according to regular and predictable changes (cf. Machamer et al. 2000: 3; Glennan 2002: S344; Bechtel 2006: 26).

Starting from this general idea of mechanism, mechanistic approaches went on to take up and develop the tradition of classical mechanical philosophy, but unlike this latter, as mentioned above, the "new mechanists" decidedly rejected a reductive

social scientist can use the ideal type of "feudalism" to investigate some cultural traits of the Incas and the Aztecs, and conclude that they do not instantiate this notion (cf. Weber 1922[1949]: 156). In this sense, the ideal type works like a centring: while it is removed as soon as the construction work is complete, that which has been built (or without metaphor, known) remains.

view of science: descriptions of mechanisms, in which mechanistic explanations essentially consist, have to be as multilevel as the phenomena they are invoked to explain.

In this connection, the possibility of interlevel explanation (and causation) is a crucial question for the new mechanistic philosophy, because it is not easy to deny that claims about interlevel explanation (and causation) are ubiquitous in scientific literature (cf. Craver and Tabery 2017, § 4.2). And from this follows that any account of multi- or interlevel explanation requires, among other things, some clear and applicable criteria for sameness of level. According to Craver's authoritative suggestion, to say that different components are on the same mechanistic level is meaningful only insofar as we assume that two conditions apply: first, components are in the same mechanism, and second, neither of them is a component of the other (Craver 2007: 191–195).

7.2.1 *Eronen's Objection*

Eronen (2013, 2015) has argued that Craver's criterion for the sameness of levels leads to a contradiction or, more precisely, to a dilemma. He observed that this definition of the property of being at the same level essentially depends on the notion of a "component". Trying to clarify the meaning of the term "component" in the writings of the chief exponents of the new mechanistic philosophy, and especially in those of Craver, Eronen distinguished a strong and a weak meaning. For our purposes, it will be sufficient to confine ourselves to the weak meaning, without any loss in the scope and validity of our conclusions.

The (weak) meaning of "component" is based on a "mutual manipulability account", which according to Craver provides a "sufficient condition for interlevel relevance": "a part is a component in a mechanism if one can change the behavior of the mechanism as a whole by intervening to change the component and one can change the behavior of the component by intervening to change the behavior of the mechanism as a whole." (Craver 2007: 141; see also Craver 2007: Ch. 4, Section 8).

This meaning of "component" leads to the consequence that "subcomponents turn out to be components of the overall mechanism." For example, by blocking the Na^+ channel of a rod cell, we change the behaviour of the overall phototransduction process or mechanism, and by exposing the retina to a light stimulus – that is, by intervening on the overall mechanism – we change the behaviour of the channel. Thus the mutual manipulability condition leads to the conclusion that subcomponents (and their subcomponents, and so on) are to be regarded as components of the mechanism as a whole (Eronen 2015: 48; Fig. 7.1).

This is exactly where a difficulty arises for Craver's criterion, which should enable us to establish whether or not different components and/or subcomponents are at the same level:

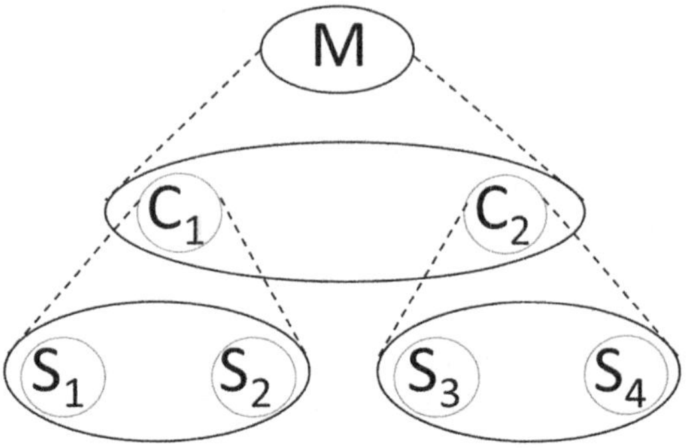

Fig. 7.1 Mechanism, components, and subcomponents: M stands for the overall mechanism, C_n are components of the mechanism, and S_n subcomponents. (Reproduced from Eronen 2015: 49)

> Let us consider [...] components C_1 and C_2 in mechanism M (Fig. 2) [here: Fig. 1, M.B.]. They are at the same level, since C_1 is not a component of C_2 and C_2 is not a component of C_1. Consider then a subcomponent S_1 of C_1. Assuming that it satisfies the mutual manipulability criterion, it is also a component of M. Furthermore, S_1 is not a component of C_2, and C_2 is not a component of S_1, so following Craver's criterion, C_2 and S_1 are also at the same level. However, if C_2 and C_1 are at the same level and C_2 and S_1 are at the same level, it follows that C_1 and S_1 are at the same level, under the very plausible assumption that the same-level relation is transitive. This leads to a contradiction, since S_1 is a component of C_1 and thus at a lower level than C_1. (Eronen 2015: 48–49)

As I have argued elsewhere, the contradiction or the dilemma only follows as long as it assumes a naïve part/whole relation between mechanism, component, and level. More precisely, the argument depends upon an atomistic interpretation of mechanism, levels and components, which (1) most of the exponents of the new mechanical philosophy, and especially Craver (see for example 2001: 67, and 2013), would not (at least explicitly) accept, and (2) which, what is more important in our context, is ruled out by a perspectival and context-relative view about mechanisms and causal attributions, according to which the (interlevel) causal interactions between a mechanism as a whole and its components, or between a mechanism and its environment, are a function of the epistemic interests of the inquirer(s).[3]

Concerning the first point, it will suffice to point out that the perspectival aspect of mechanisms (though with some vacillation: see Buzzoni 2016), is not ignored by the leading exponents of the new mechanistic philosophy (cf. Darden 2008: 959; Craver 2007). Craver 2013, in particular, has stressed the dependency of mechanisms from the pragmatic aims of the researchers (and mitigated this dependency only by James Woodward's interventionist approach to causality). For example, he holds that, although levels of mechanisms satisfy many of the main features associated with levels in the first place, they

> fail to satisfy other widely held beliefs about levels. First, in contrast to the common way of speaking about levels, levels of mechanisms should not be conceived as levels of objects (for example, societies, organisms, cells, molecules, and atoms). They are levels of behaving components. In many cases, the components picked out in a mechanistic decomposition fail to correspond to paradigmatic entities with clear spatial boundaries. [...] Second, unlike

[3] As far as the second point is concerned, my reply to Eronen's objection draws on Bertolaso and Buzzoni 2017, especially § 3.

> Oppenheim and Putnam's six levels of nature, levels of mechanisms are not monolithic divisions in the structure of the world. (Craver 2007: 190)

The last sentence quoted leads to the second point, which is more important for our purposes. It is interesting to note that Craver, one page later, commenting on a diagram by Wimsatt, which is similar to that of Eronen, writes as follows:

> levels of mechanisms are far more local than the monolithic image [sc.: in Wimsatt's diagram] suggests. They are defined only within a given compositional hierarchy. [...] How many levels there are, and which levels are included, are questions to be answered on a case-by-case basis by discovering which components at which size scales are explanatorily relevant for a given phenomenon. They cannot be read off a menu of levels in advance.[4]

Apart from questionable details (such as the reference to size scales), into which I cannot enter now, this is essentially the point I want to urge. More precisely, it can be said that the atomistic interpretation of mechanisms, components, and subcomponents, which is implicitly assumed in Eronen's argument, is ruled out by a perspectival and context-relative view about mechanisms. Thus, the soundness of the argument is undermined by the problematic assumption of an atomistic or discrete account *not only of components and levels, but also of mechanisms*, that is, by the assumption that both the components of a mechanism and the mechanism itself retain their distinct self-identity, whatever be the context into which they enter.

In the case at hand, according to Eronen, the contradiction arises from the fact that the components S_1 and C_2 cannot be said to be and not to be on the same level. But if the notions of mechanism, level and component are relative to the choice of a theoretical perspective that predetermines, at least in a *generic* sense, what may be regarded as a suitable mechanism, level or component for the investigation of a particular subject matter, the contradiction only follows if there is no shift in the meaning of the terms involved. As is well known, there is a contradiction only if we affirm and deny anything of something "in one and the same respect" (cf. Aristotle, *Metaph.*, IV, 3), that is, as regarded in one and the same sense. But it is exactly this condition that is called into question by the context dependence of mechanisms, levels, and components. It follows that, in a sense, S_1 and C_2 *are* on the same level if this is taken to mean that they are components of the same mechanism M that fulfil the mutual manipulability condition; they are on the same level from the point of view of the same mechanism or, more precisely, *from the vantage point of the theory which describes that mechanism*; however, in another sense, they might be regarded as not being on the same level, for example *if they were concepts of a*

[4] Craver 2007: 191. In order to prevent the misunderstanding that might arise from taking this quotation out of the context, it should be said that Wimsatt's diagram is to be regarded, in the last analysis, as only an idealized model, and his view is more similar to that of Caver than would appear at first sight. When we come to read Wimsatt, it is always important to pay attention to the fact that, in a way very similar to that of Thomas Kuhn, he hedges his formulations with many restrictions and limiting clauses. For example, concerning the point in question, Wimsatt writes: "compositional levels of organization are the simplest general and large-scale structures for the organization of matter. They are constituted by families of entities *usually* of comparable size and dynamical properties, which characteristically interact *primarily* with one another, and which, taken together, give an apparent rough closure over a range of phenomena and regularities." (Wimsatt 2007: 204; my italics).

theory that distinguishes sharply between two different mechanisms, of which S_1 and C_2 are respectively regarded as 'parts'.

In other words, the question concerning the sameness of mechanism or level cannot be answered without a perspectival approach. The question as to what mechanisms and levels are cannot be answered in a purely abstract way, but only with a definite aim in view. Given the intimate connection between mechanisms and experimental setups, this point may be explained with the help of the similar problem concerning the possibility of re-identifying experiments. When can we speak of the 'same' experiment? Galileo's original experiment of a free falling body and its present repetition performed by first year physics students using very sophisticated devices (such as photoelectric cells, electromagnets, digital clocks) are, despite the different devices, the same experiment because *each brings to light the same 'mechanism', that is, the same real causal connection that constitutes a specific answer to the same theoretical question.* And conversely, by changing the theoretical framework two instances of one 'same' experiment can become two different experiments. Two experiments, identical as to the experimenter's actions and the experimental mechanism, can stand for two distinct experiments, or even two experiments in distinct scientific disciplines, if performed from distinct perspectives, in order to answer distinct theoretical questions. Before 1905, experiments on the composition of velocities were considered the 'same' whether the velocities they involved were far from or close to the velocity of light, since Newtonian mechanics does not distinguish on this basis. After that date, in the light of the special theory of relativity, these experiments take on entirely different meanings and therefore should be considered different experiments (for this example see Franklin 1989: 438–439; see also Nickles 1988: 33–38).

For similar reasons, by changing the theoretical framework two instances of the 'same' mechanism can become two different mechanisms if their reproducibility and their "regular changes from start or set-up to finish or termination conditions" (Machamer et al. 2000: 1) are no longer identical. And what may appear as a mechanism in a particular causal context may be regarded as a level, or even a component, in another theoretical framework (and *vice versa*). (For a more detailed discussion of this point, see Buzzoni 2008, chapter 1, § 3).

It follows from this that a naive, and more precisely, an explicitly or implicitly atomistic conception of what "mechanisms", "levels", "parts", or "components" of a mechanism are, is untenable. Eronen's dilemma follows only on the untenable assumption that the individual components of different mechanisms do not change their nature when taken into new mechanisms, that is, on the assumption that, although new combinations of components may arise, they remain the same identical components. If context-relativity is admitted, we may avoid the contradiction. In this case, Eronen's objection presents a difficulty for Craver and other mechanist philosophers, *only if, and so long as,* they assume an atomistic notion of mechanism, level, and component. Explanations may be given upon different levels, but levels cannot be known independently of the choice of a particular perspective, whose change carries with it change of (ontological) level and (methodological) explanation.

7.2.2 Two Difficulties

Many other scholars have taken a similar path by stressing the context relativity of mechanisms and levels and by putting them in relation to pragmatic factors (see for example Strand and Oftedal 2009; Potochnik and McGill 2011; Pâslaru 2009; Rueger and McGivem 2010, Bitbol 2014). There are, however, at least two connected difficulties still to overcome. First, if context dependency is not sufficiently well understood, we easily fall into the (ontological) error of undermining the objective side of scientific explanation. Second, though a perspectival conception according to which different perspectives lead to different notions of mechanisms, levels and components, is a necessary step towards a satisfactory account of interlevel interaction and explanation, it is not sufficient to understand how different perspectives may interact or clash with one another when addressing the same problem.

There is no space in the present paper to examine the first point at the length that it deserves, and I shall confine myself to the brief discussion of one example (for a more detailed analysis, see Buzzoni 2016). The context relativity of mechanisms and levels is maintained by Bitbol 2014 on the basis of an "interventionist-constitutive conception of causation", which draws largely on Kant. In his *Prolegomena,* Kant maintained that spatial and kinematic predicates can (and should) be considered "secondary" in Locke's sense because they are relative to the pure form of our *sensibility.* But, according to Bitbol, by an extension of this same point of view he could also have said that ascribing the status of a "substance" (or the status of a "cause") to something is to be regarded as secondary in the same sense, "because both ascriptions are relative to the pure form of our *understanding.*" (Bitbol 2014: 694)

To maintain context relativity of mechanisms and levels on this basis means to undermine the objective side of scientific explanation, even if one combines context relativity with an objective theory of causality. On the one hand, Bitbol rightly insists that any causal scheme is relative to "the method of active substitution of antecedent by means of various instruments adapted to various scales or levels." On the other hand, however, the most important consequence of this Kantian approach is that, if each model or level is relative to a completely distinct class of modes of access and modes of conceptualization, it is nonsensical to posit real interactions between entities of two different models or levels of a certain process: as Bitbol notices, "[n]o level of organization can claim any privilege for itself, because every such level is defined (or "constituted") by a certain scale of intervention and observation." (Bitbol 2014: 694–695)

It is easily seen that this last claim implicitly denies the objective value of mechanisms and levels: known levels do not possess any kind of independent existence apart from human intervention, and thus they turn out to be only an artificial creation. According to Bitbol, levels are constructed (and not *reconstructed*, as it should be) by our causal interventions on them. In this way, the refusal to distinguish between phenomenon and thing in itself (a point on which I agree with him) is pushed to the point (which I cannot accept) of obliterating or even blurring the distinction between knowledge and the object of knowledge; for, were this distinction suppressed, there would be no reason why levels ever should have come to make up the one common world of human scientific inquiry.

And this leads to the second point, much more important for present purposes. If each mechanism, level or component is relative to a completely distinct perspective or mode of conceptualization, it is difficult to see how there can be room for alternative explanations that are provided in the terms of entities belonging to different and autonomous levels.

This point is important for our present purpose as it is a serious challenge not only to Bitbol's position, but also to the perspectival view of mechanisms and levels assumed here. In order to meet this challenge we need to set it in a wider context, following Wimsatt 1974, who was among the first to raise the question of interlevel explanation in biology in connection not only with mechanisms, but also with the problem, taken in its widest sense, of intertheoretic reduction.

Contrary to what is often assumed, an acceptable solution to the question of the nature of *interlevel* interactions and explanations cannot be given separately from that of the old and difficult problem concerning *intertheoretic* reduction (and *vice versa*). This already follows from the perspectival point of view here adopted: if different mechanisms, levels, and components are differently selected by different theoretical perspectives, one cannot answer the question concerning the relations between levels without answering the question concerning the relations between different or even competing perspectives and theories (and *vice versa*). But it is also a fact that according to important exponents of the new mechanistic philosophy, the question of interlevel explanation and intertheoretic reduction are intimately connected. On the one hand, they introduced mechanisms as a key concept to help resolve traditional epistemological problems, among which Machamer, Darden and Craver included that of understanding the nature of intertheoretic reduction:

> Theory change in neuroscience and molecular biology is most accurately characterized in terms of the gradual and piecemeal construction, evaluation and revision of multi-level mechanism schemata [...] Elimination or replacement should be understood in terms of the reconceptualization or abandonment of the phenomenon to be explained, of a proposed mechanism schema, or of its purported components. This contrasts with the static two-place relations between different theories (or levels) and with the case of logical deduction. (Machamer et al. 2000: 23).

More particularly, Craver 2007, to whom we owe fundamental clarifications on the relationship between interlevel explanation and intertheoretic reduction, admits that *reduction* models involve "relationships between theories *at different levels*" (Craver 2007: 256; italics added).

In other words, as long as the concept of interlevel explanation presupposes the concept of relationships between *theories* at different levels, a clarification of interlevel explanation involves a clarification of the relationships between the various theories that scientists wish to work with, a relationship that is the common basis of all specific theories of reduction in science. This is what will occupy us in the last part of the paper. The classic problem concerning intertheoretic reduction is too complex and many-sided to be examined at the length that it deserves, but we may indicate the direction in which an acceptable solution might be found.

7.3 Distinguishing 'Weak' and 'Strong' Relations Between Knowledge Claims

For this purpose, I shall approach the problem by distinguishing, provisionally and ideal-typically, two sorts of relations (be they intertheoretic or interlevel) between knowledge claims: *weak* and *strong*. As we shall see in the rest of the paper, this distinction is provisional in a sense similar to that in which a masonry falsework is useful as a temporary support, but may be removed when the work is complete. Both types of relation are only ideal types between which we find an indefinite number of intermediate forms of relations. Three further connected points will be made after the distinction is introduced: (1) The two kinds of relation between theories or levels have in common that relating multiple perspectives to one another in order to better understand the subject-matter under investigation requires constructing a new, wider or deeper perspective. (2) The construction of a new, wider or deeper perspective is embedded in a historical and practical context and cannot be devised philosophically (even by philosophers of science). It can only be the product of working scientists, which specify reduction relations that are themselves a dynamic element of the historical development of science. (3) This is true not only for physics, but for all empirical sciences (Sects. 7.4, 7.5, 7.6 and 7.7).

The first sort of intertheoretic or interlevel relation, the weak one, holds when explanations offered from different and autonomous perspectives contribute to understanding the same phenomenon from different and autonomous viewpoints.[5] In this case, not only is there no a priori reason for preferring one theory or level to another, but they are at least prima facie fully compatible with one another, so that one can safely adopt both of them without contradicting him or herself. If two people look at the same thing from different positions, the sentences describing the different aspects of the thing can be combined without contradiction under the condition that they are conceived as referring only to the corresponding aspect. In the same sense, it would seem, at least at first sight, that propositions of different disciplines about the same thing are not in contradiction with one another: "This watch is made of gold", and "The value of this watch may be estimated at $500" are loosely connected *a parte rei* and are not in contradiction with each other (for a similar example, see Agazzi 2014: 83).

A *prima facie* very different kind of intertheoretic or interlevel relation, which I shall call *strong*, is that which holds between explanations in which theories have

[5] That explanation increases *understanding* is a much disputed issue today (cf. for example De Regt and Dieks 2005), to which the effort to grasp the explanatory power of thought experiments in science provided an important contribution (on this point see above all Stuart 2016 and 2018, where further references will be found).

overlapping domains of application and are for this reason in competition with one another. In the preceding example of two people looking at the same thing, this is the case if they are looking from the same position, but describe different aspects of the thing by sentences that are, for whatever reason, incompatible. Often, scientists formulate this kind of intertheoretic relation in the course of attempting to increase the epistemic quality and/or explanatory power of a theory by clarifying its relation to previous theories, especially those with competing solutions for the same problems. It is this sort of intertheoretic relation (usually called "reduction") that logical empiricists considered the most distinctive mark of scientific progress, because theories supersede and replace rival theories that are later derived as "special cases". Another example of this relation in philosophy of science can be found in Imre Lakatos's and Larry Laudan's answers to Kuhn and Feyerabend's challenge, as they assume that this kind of intertheoretic relation must be possible if we are to speak of scientific progress.[6]

7.4 Weak Intertheoretic Relations Between Knowledge Claims

Let us briefly examine these two kinds of interlevel or intertheoretic relations. In the vast literature dealing with mechanistic philosophy, it was Craver who most clearly expressed the first sort of interlevel or intertheoretic relation, with specific reference to neuroscience. Craver writes:

> whereas reduction models involve global relationships between theories at different levels, the mosaic model accommodates the fact that interlevel relations are often formulated piecemeal, within local mechanisms, by adding constraints on interlevel relations. (Craver 2007: 256)

The different fields that contribute to the "mosaic unity" of neuroscience, though providing some constraints on a mechanistic explanation, "are autonomous in that they have different central problems, use different techniques, have different theoretical vocabularies, and make different background assumptions" (Craver 2007: 231; all of his chapter 7 is devoted to the "mosaic unity of neuroscience")

Craver's concept of the mosaic unity of neuroscience captures an important feature of 'weak' relations between sentences or theories. Using it, Craver downplays intertheoretic reductions to the point of making them irrelevant for biology. In this

[6] As the following discussion will show, the distinction between "strong" and "weak" reduction, despite the terminological identity, is very different from those already suggested by Kimbrough (1979: 394–395) and Primas (1985: 109). Apart from its provisional and ideal-typical character, my distinction is much more general: in both cases, "strong" and "weak" reductions fall under my category of "strong" relations between knowledge claims. Primas' and Kimbrough's use of these expressions corresponds roughly to the usual distinction between a direct or derivational kind of intertheoretic reduction and an approximate or indirect one, of which I shall say something in the next section.

way, he attempts to free himself from Nagel's classical model of reduction, which, according to his own definition, would be achieved "by identifying the kind of terms in higher-level theories with those of lower-level theories and deriving the higher-level theories from the lower-level theories." For Craver, reduction in this sense is so "peripheral to the practice of neuroscience" that it is even "misleading" to conceive of it "as a regulative ideal for integrating neuroscience" (Craver 2007: 17–18).

7.4.1 Two Kinds of Weak Intertheoretic Relations

For purposes of exposition, one might distinguish two main cases of weak (interlevel or intertheoretic) relation, with the proviso that they are often so closely interwoven with one another in scientific practice that it is very difficult to disentangle them. The first one is when explanations are offered from different and well-established scientific disciplines like physics, chemistry, or biology. Scientific disciplines investigate the same reality from different perspectives and cast light on different aspects of it, and intertheoretic relations help us both to articulate the position of each particular scientific discipline vis-a-vis the others, and express each's relative autonomy. For example, a mechanical phenomenon results from considering reality from a partial perspective, which takes into account only some properties of reality, such as force, mass and certain spatial and temporal relations. This relativization of any scientific discipline to the choice of a perspective on reality vindicates its autonomy from any other scientific discipline, since there is no a priori reason for choosing one perspective rather than another. A table can be examined as a physical phenomenon as well as a chemical one, and there is no a priori reason to adopt one perspective rather than the other.

A second class of cases in which the weak kind of (interlevel or intertheoretic) relation obtains is between sentences or theories that, though they belong to the same scientific discipline, explain something in terms of different and *prima facie* autonomous concepts, instruments and techniques (and therefore, from the mechanistic point of view, in terms of different and *prima facie* autonomous mechanisms, levels of organization, or components). Although theories belonging to the same discipline are much more closely related to one another than theories belonging to different disciplines, they do not lead, more often than not, to mutually incompatible claims, and for this reason they are connected in such a way that they may be used together to solve the same problems.

Take for example protein folding, which is the process by which a protein assumes its characteristic structure, known as the native state. The most important question here is how an amino acid sequence specifies both a native structure and the pathway to attain this state. As Chen et al. 2008 have shown, the field of protein folding developed from three different perspectives: the first is the physics of intramolecular forces modelled *in silico* (that is, performed on computer or via computer simulation), the second is the experimental study *in vitro* (that is, in a controlled environment outside of living organisms), which is based on techniques of structural chemistry, and the third is the *in vivo* analysis of positions, interactions and functions, using techniques of cell biology (Chen et al. 2008). The same phenomenon of protein folding may be looked at from different and relatively autonomous standpoints. In the first case, physical levels, causes and effects are in the foreground; in the second, chemical explanations; in the third, cell biology.

Protein folding is a good instance both of weak intertheoretic or interlevel relations, and of the fact that the two kinds I have just distinguished are often closely interwoven. On the one hand, the same protein folding has been looked at from different, but relatively autonomous disciplinary standpoints. On the other hand, however, this example is essentially taken from biology, interpreted here in the simple but historically correct sense as the scientific study of living things.

7.4.2 Scientific Progress

In general, speaking of scientific progress does not make sense in reference to weak intertheoretic relations, because these different perspectives are (*prima facie* at least) not "superior" to one another. To the extent that different perspectives are seen as incommensurable and independent ways of describing the same reality, each makes its own contribution to understanding that reality. On closer examination, however, weak intertheoretic relations are not really as weak as they may initially appear, for the following reasons.

First, they contribute to progress in at least one very obvious way. Even in the case of purely incommensurable and independent perspectives, the opening of a new perspective from which to describe reality widens and enriches our understanding of the world.

Second (and most importantly for our purposes, though much more difficult to express clearly) scientists are driven from narrower perspectives to wider and deeper ones by some aspects of reality that retroacted and resisted reduction not only to any particular perspective, but also to the co-ordination and integration of different but compatible perspectives. We can view reality from different perspectives, but what we observe has its content independently of the perspective that was used to discover it and leads science as a whole to pursue the ideal of ever-increasing rational coherence and systematic unity. To say the same thing from a slightly different point of view, the regulative ideal of integrating or unifying all possible empirical perspectives is based on the fundamental assumption of science that empirical reality, to which in the last analysis all scientific statements are supposed to refer, can only develop in one way. For this reason, it is in principle not only possible (as for example O'Malley et al. 2014 points out), but also opportune and desirable, even in the same scientific discipline, to approach problems from multiple perspectives. Both in the case of *prima facie* compatible perspectives and in that of clearly different scientific disciplines, scientists should be driven by the regulative ideal, so to speak, to achieve a complete coherence in our experience, and this can only be done, though always provisionally, by incorporating different perspectives into wider, more inclusive perspectives.

To make this point a bit clearer, let us reconsider the case of protein folding. I said before that research concerning protein folding seems to contain examples of weak intertheoretic relations as it shows the explanatory and ontological relativity of different perspectives and levels. Although this is relatively accurate, a more careful examination shows us that, in this case too, it is not difficult to find a second thread in protein folding, which, though not ultimately inconsistent with the above analysis, tends in the direction of integrating – and, as a matter of fact, it is beginning to integrate – the physical, chemical and biological perspectives on protein folding.

Protein folding research, as already mentioned, passed through different phases. More precisely, according to Chen et al. 2008, during the first phase, research was essentially pursued through *in silico* and *in vitro* studies, while the second phase in protein folding research was defined by questions concerning the influence of other biological molecules in the cellular environment. Now, *in vivo* studies are much more complex than *in silico* or *in vitro* ones. But it is exactly this characteristic of complexity that requires the emergence of a broader perspective, which enables us to assess the influence of the cellular environment on protein folding. As Chen et al. 2008 note, all of the physical and chemical considerations concerning protein folding still apply to some degree to the study of aggregating proteins. But it is evident that, as these scholars partly explicitly and partly tacitly recognize, the study of protein folding in the context of external influences like protein concentration, localization, or evolution is a new and broader perspective, which cannot be reduced to the simple application of knowledge already gained by *in silico* and *in vitro* methods. On the one side, the discussion has been driven back into its native environment, i.e. inside living organisms, but on the other side this has involved a partial change and an important extension of its original meaning:

> As this field develops and knowledge of protein aggregation as a general phenomenon accumulates, we stand to gain not only vital tools for treating specific diseases, but also insight into the behavior of all proteins with respect to their environment. (Chen et al. 2008, § 7)

From this point of view, the *in vivo* perspective is a new and wider perspective, which at the same time shows some limits of the *in silico* and *in vitro* points of view. We are thus in a position to conclude that weak intertheoretic relations not only do not exclude progress, but they always imply a certain tension between the co-operating perspectives or theories, a tension which points in the direction of more inclusive perspectives, and therefore of a kind of progress in the sense of the 'strong' kind of intertheoretic relation.

7.5 Strong Intertheoretic Relations Between Knowledge Claims

In the preceding section I pointed out that, contrary to what may have been suggested by our initial definition, weak intertheoretic relations are not actually weak in the sense of being ineffectual for scientific progress. In this section, I shall argue that we are led to a similar, though reversed, conclusion if we examine the 'strong' intertheoretic relations (among which are those usually called intertheoretic reductions), in which theories are in competition because they provide alternative explanations of the same phenomena. We shall see that they are not really as strong as they initially appear. And this will put us in a position to extend our conclusions from intertheoretical to interlevel explanatory relations or reductions. Furthermore, it will emerge that there is no reason in principle why interlevel relations that are to some extent strong should be excluded from scientific fields other than physics (Sects. 7.6 and 7.7).

7.5.1 Two Kinds of Strong Intertheoretic Relations

Within the general framework of logical empiricism, a long-standing debate soon led to the distinction between two kinds of strong intertheoretic reduction, that is, between a direct one and an approximate or indirect one. The first kind is the typically derivational reduction that is usually attributed to Nagel: the basic terms (and entities) of one theory are related to the basic terms (and entities) of the other, and the axioms and laws of the reduced theory (for example, Galileo's laws of freely falling terrestrial bodies) are derivable "as a special case" from the reducing theory (for example, the Newtonian theory; cf. Nagel 1949: 103; more generally see Nagel 1961, and Kemeny and Oppenheim 1956).

The second kind of strong intertheoretic reduction is exemplified by the more general schema proposed by Schaffner: a theory T_1 corrects a theory T_2 in the sense of providing "more accurate experimentally verifiable predictions than T_2 in almost all cases, and should also indicate why T_2 was incorrect (e.g., crucial variable ignored), and why it worked as well as it did."[7]

As an example we may take classical and relativistic momentum. Relativistic momentum is expressed as follows:

[7] Schaffner 1967: 144. For more recent developments of this idea, see above all Schaffner 2012. Among the many authors who have taken up (though with some modification) Schaffner's earlier model of approximate reduction, see for example Goosens 1978, Dizadji-Bahmani et al. 2010, Klein 2009, and Needham 2010. Another important but little recognized contribution to intertheoretic reduction in this sense was given by Erhard Scheibe (cf. Scheibe 1997, 1999; Scheibe 2001: Part V), who developed a theory of reduction which, at least in the sense that it is constructive and integrative, has some points of contact with the view expressed here.

$$p \equiv \frac{mv}{\sqrt{1 - v^2 / c^2}}$$

Where v is the velocity, say, of some particle, and m is its mass as measured by an observer at rest with respect to the particle. If v is much smaller than c, the denominator approaches 1 and p approaches mv. In this sense, classical equations become a special case of relativistic equations (or, as scientists usually say, relativistic equations reduce to classical equations; on this point, see Nickles 1973).

In 1957 Popper raised a strong objection against Nagel's idea of reduction, elaborating a point already made by Duhem in his *La Théorie Physique* (cf. Duhem 1914, chapter 6, § 4). From a logical point of view, there can be no relation of deducibility between Newton's theory and Galileo's, since Newton's theory contradicts it and no conclusion that contradicts the premises can be validly inferred (cf. Popper 1972: 199–201; cf. also the similar point made later by Scheibe 1973[2001]). Contrary to what Galilei believed, according to Newton's theory the acceleration of free-falling bodies is never constant, and to say that Galilei's equations can yield good approximations from the point of view of Newton's theory if the height is negligible as compared with the radius of the Earth, is "strictly speaking false" (Popper 1972: 201). Hence, according to Popper, we may conclude that, "[f]ar from repeating the *explicandum*, the new theory contradicts it, and corrects it." (Popper 1972: 202)

This argument is probably valid against Nagel's derivational view of reduction (which however is not as simple as is often supposed: cf. Needham 2010), but only half refutes Schaffner's approximate or indirect kind of reduction. The fact that, as Popper says, the new theory contradicts and corrects the older one, so that, as Kuhn put it, the older theory can be regarded as a special case only after being "transformed for the purpose" under "the explicit guidance of the more recent theory" (Kuhn 1970: 103), is less a refutation of Schaffner's model (or the many varieties of it, some of which I have cited in footnote 6) than a reformulation to meet the objections urged against Nagel's account.

According to Kuhn, that the previous theory is corrected under the guidance of the more recent theory implies that science does not move towards a true representation of the way the world is in itself. But to accept Kuhn's premise does not compel us, without any other supposition, to accept Kuhn's conclusion. Interpreted differently, this point only expresses an important element of truth in Schaffner's model, which could be expressed as a sort of asymmetry between verification and falsification, *an asymmetry that is not-Popperian* because it expresses a view of scientific progress that is diametrically opposed to Popper's conception. It is true, as Popper pointed out, that the relativistic correction ($\frac{1}{\sqrt{1 - v^2 / c^2}}$) necessary to deduce one equation of motion from another concerning the speed of the particles, is the most important element in the real progress made by Einstein's theory. But the essential point (which Popper cannot consistently admit because of his falsificationist view) is that Newton's theory *had, and ought to have, the means of establishing its own truth* (or—because it does not matter for my general argument—its own verification, confirmation, corroboration, empirical adequacy, reliability, etc.), *while Newton's theory did not have, and could not have, the means of establishing its own falsity, or, better, the limits beyond which, if exceeded, it would lead to false (or unconfirmed, empirically inadequate, unreliable, etc.) predictions and to a loss of technical control over reality*. Let us briefly discuss both parts of this last claim.

On the one hand, Newton's theory had, and ought to have, the means of establishing its own truth. As any other scientific theory, it was based at least on some predictive, explanatory, and technical-experimental success concerning one or more fields of experience. This is in itself a fairly obvious point, but it is intimately connected with the usual and interesting practice of physics to quantitatively confirm or corroborate new and more comprehensive theories, such as relativistic and quantum mechanics in comparison to Newton's mechanics, by assuming the truth of the old theory, that is, of the theory of which the more accurate predictions of relativistic and quantum mechanics are intended to show the limits.

In order to argue for the absurdity of traditional methodology and the fruitfulness of counterinduction (that is, of the rule advising us "to introduce and elaborate hypotheses which are inconsistent with well-established theories and/or well-established facts"; Feyerabend 1993: 20), Feyerabend noticed that the relativistic calculation of the path of Mercury is a "veritable nightmare" because it uses, and must use, classical physics "up to a certain point (which is often quite arbitrary)" and adds the new theory "for calculating refinements." (Feyerabend 1993: 47–48) However, from this fact we may draw a lesson which is very different from Feyerabend's methodological anarchism, namely, that what is true (empirically adequate, robust, reliable, or whatever) in relativistic physics cannot be justified without recognizing the approximate truth (empirical adequacy, etc.) of Newton's theory of gravitation.[8]

Strictly speaking, therefore, it is not true that Einstein's relativity theory can do everything the classical theory could, as authors as different as Nagel and Feyerabend have implicitly or explicitly accepted (cf. Feyerabend 1966: 244). On the contrary, *we need to assume that Newton's theory is at least approximately true (empirically adequate, reliable, etc.), in order to test Einstein's theory.* Newton's mechanics is in an important sense "closed" or self-contained, or indifferent to the relativistic

[8] For a similar remark, see Krüger (1974: 16) and Scheibe (1973, 2001: Part V). More recently, a similar point has been made by Falkenburg 2007 and Gutschmidt 2014, §§ 2.1 and 2.2. In particular, Brigitte Falkenburg has shown this point very clearly for the measurement methods of particle physics, and more precisely for the quantitative analysis of curved particle tracks obtained in a bubble chamber in a magnetic field (see Falkenburg 2007, for example pp. 96–119).

effects, in the sense that it does not require Einstein's theory (or any other *successive* theory) to be understood and empirically tested.[9]

On the other hand, Newton's theory did not have, and could not have, the means of establishing its own falsity. Feyerabend is surely right in pointing out, with the great majority of philosophers, that the theory of relativity "can do a few more things" than the classical theory could, and that "it is successful where classical physics fails" (Feyerabend 1966: 244). For, as already mentioned, although Newton needed some reason for asserting his theory, strictly speaking he could not have the means of determining its limits. The limits within which a scientific theory appears to be true can be known only indirectly, through the failed efforts to successfully apply the theory to a field of experience that should fall under its domain but does not. For instance, while the well-known anomaly in the perihelion advance of Mercury's orbit was certainly an anomaly or at least an open question at the time of Newton, it could not be considered as a general or theoretical limit from the point of view of Newton's celestial mechanics; from that point of view, the anomaly had to be supposed to be in principle eliminable. In order to determine the limits within which a theory may successfully be employed, scientists require, explicitly or implicitly, a new theory: one can see errors or inadequacies of an earlier perspective only from a new perspective. In order to know directly the limits of a scientific theory, we have to overcome them, as happens when scientists find a new theory that is (relatively) more accurate and comprehensive. Newton probably never seriously envisaged the possibility that anything can move with a velocity greater than that of light, or that the laws of classical mechanics do not work at the atomic scale. But, even if he had done so, he would have lacked all the necessary conceptual and/or technical instruments in order to assign an empirical meaning to his hypotheses.

If we subscribe to the thesis of the intrinsic unity of theory and technique in science, and apply it to our problem, we may express both parts of our main claim by saying that the knowledge of the world made possible by Einstein's theory does not and cannot render obsolete the knowledge made possible by Newton's theory, since Einstein's relativity theory (or quantum mechanics) does not and cannot deny the technical accomplishments of Newton's mechanics, but, on the contrary, starts from them, and then comes to a more comprehensive because more differentiated experimental (or, what is the same thing, technical) control over reality. In the change from Newtonian to Relativistic dynamics and kinematics, greater refinements or discriminations in semantic (extensional and intensional) level go hand in hand with greater experimental-technical discrimination: Einstein's theory distinguishes and discriminates between experimental-technical settings that cannot be distinguished

[9] Erhard Scheibe (1993[2001]: 136–141) aptly drew the attention of the philosophers of science to Heisenberg's and von Weizsäcker's concepts of "closed" and "self-contained" theory (cf. for example Heisenberg 1959: 78, and von Weizsäcker 1971: 213 ff.). Though having a reference to Heisenberg's and von Weizsäcker's corresponding expressions, the words "closed" and "self-contained" as here used must be interpreted in the present context, that is, contrary at least to von Weizsäcker's definition, in such a way as to admit changes in the theory that may in principle affect both laws and concepts without necessarily obliging the scientists to say they are working with a different theory.

or discriminated by Newton's theory. The meaning of the term "mass" in classical physics is relatively less determinate, in the sense that it fails to discriminate semantic and experimental-technical aspects that should not be confused with each other from the viewpoint of Einstein's theory: for this reason, Newton's theory leads to false conclusions only if applied to physical situations which it cannot discriminate by its experimental-technical concepts and methods.

There is not enough space to develop this view presently at the length that it deserves.[10] But, as I shall explain in the remainder of this section, the preceding considerations should be sufficient to make clear that strong intertheoretic relations are not really as strong as they may initially appear.

As already mentioned, the deduction (in the asymptotic limit) of one theory as a special case of another — which perhaps best instantiates the strong kind of intertheoretic reduction — can be expressed by the deductive-nomological model of explanation only after having corrected the first theory under the guidance of the second. But it is important to notice that here the genuine scientific progress (that is, the progress in terms both of new experimental-technical applications and the specification of the boundaries of the old theory beyond which it turns out to be false, unreliable, falsified, etc.) *consists exactly in the corrections made by the theory of relativity of Newton's laws in order to state a relation of deducibility between the two theories.* And it is equally important to stress that these corrections, if we want to reliably speak of progress in science, must pass certain crucial experimental tests. It follows, on reflection, that deduction (in the asymptotic limit) of, say, Newton's laws as a special case of the relativistic ones, taken in itself, proves very little (and strictly speaking, nothing at all) about the *empirical* truth of Einstein's theory (and about the limits of Newton's) apart from *some experimental results that are empirically regarded as sufficiently well-established and reliable.*

In other words, it is tautological to say that Einstein's theory reduces Newton's *if one neglects the experimental, and at the same time technical successes of Einstein's theory* (or, as Popper and Lakatos would perhaps put it, its new corroborated empirical content). But these successes have to be taken into account and ascertained in each particular case. For this reason, it is rather difficult to establish that a purported intertheoretic reduction in the empirical sciences is a genuine, actual, or sound instance of reduction. Empirical and experimental successes are affected by the complexity of the relationship between theory and experience. This complexity has many forms: to mention only one example, it depends on the well-known problems that Duhem raised and that we find, revived and brought up to date, in the work of Quine. Thus, even the 'strongest' intertheoretic reductions are *de facto* always problematic *for empirical and experimental reasons*: no intertheoretic reduction can be stronger than its weakest (empirical and historically conditioned) link, here represented by the successes of the reducing theory, which are, at least in in principle, always exposed to the risk of being refuted by new experience.

[10] For more details on this point, see Buzzoni 1995. It seems to me that the core of this thesis has not been rendered obsolete by subsequent inquiry, though many noteworthy contributions to this issue have appeared in the meantime.

This indirectly supports Schaffner's idea that models of reductions, even those which are usually considered the strongest, are only regulative ideals that guide the search for explanations. It has been pointed out by some authors that a strong intertheoretic reduction, which could be put in a clear mathematical form, is *de facto* a rare scientific achievement not only in contemporary neuroscience, but even in physics (see for example Schaffner 2012: 550; Hooker 2004: 437; Batterman 2002: 95). And in this sense it is important to say with Thomas Nickles that, although it is true that intertheoretic reductions involve the relation of a theory to their special cases, there are many different ways one theory may be a special case of another. Besides reduction in the sense of Nagel and Schaffner, there is "a varied collection of intertheoretic relations rather than a single, distinctive logical or mathematical relation".[11] However, the formulations of these claims need to be radicalized in principle: in the empirical sciences, a strong reduction in its pure form, such as those expressed in a clear mathematical form, is only an ideal limit, *which is really never reached in practice.*

Schaffner himself finally came to regard his model of reduction as only a "regulative" ideal that guides the search for explanations, and admitted the possibility of a "continuum of reduction relations", in which not only a completely integral theory but also portions of a theory associated with an experimental subject area or domain are preserved and modified in the context of a successive theory (see for example Schaffner 1977, 1992: 319–320, and 2012: 535). This is a very important step towards a view that incorporates the critiques of Popper, Feyerabend, and Kuhn into a more pragmatic model of reduction. But once that first step has been taken it should be consistently developed and followed to its ultimate consequences.

Moreover, and in accordance with our present view, Schaffner 2012 has emphasized again the importance of his old notion of "analogy" in connection with this issue. According to Schaffner 1967, one of the necessary conditions that the general paradigm of reduction had to meet was as follows: "(5) The relations between T^2 [*sc.*: the secondary theory] and T^{2*} [*sc.*: the *corrected* secondary theory] should be one of strong analogy—that is (in current jargon) they possess a large 'positive analogy.'" (Schaffner 1967: 141) It is already clear from the quoted passage that Schaffner is undoubtedly influenced by Mary Hesse's view, but he never sufficiently clarified his idea of analogy. But this notwithstanding, it is worth remarking that Schaffner (2012) refused to abandon this notion, even though it had repeatedly been a target of criticism; on the contrary, he explicitly emphasised again the importance of this notion of "analogy", and against Winther 2009 he noted that this author had

[11] Nickles 1973: 181. A similar idea, though only partially, is also a characteristic of the "new wave" model of reduction; see especially Churchland 1979 and 1984, Hooker 1981, Bickle 1998 and 2003. In spite of some wavering in their analysis of the logical nature of reduction, which essentially consists in deduction (cf. Churchland 1979: 81), the "new wave" model of reduction owes to Nickle the idea that "we must be prepared to count reducibility as a matter of degree. Like translation, which may be faithful or lame, reduction may be smooth, or bumpy, or anywhere in between" (Churchland 1979: 84). For some critical comments on Bickle and the "new wave" model of reduction in science, see Endicott 1998 and 2001.

just misinterpreted his original idea of analogy as needing "a formal rendition." (Schaffner 2012: 545)

It would be interesting, but it is not the place, to examine the idea of analogy in Schaffner's work with respect to the experimental-operational approach assumed here. It is enough to urge here that, in the last analysis, Newton's and Einstein's "mass," though importantly different, are strongly analogous in some respects because we know in what circumstances and under what conditions they lead to the same experimental results. This is exactly what we should expect if reduction is, as we have suggested, intimately connected with experiment. Though strong intertheoretical reductions may always, at least in principle, be expressed as logico-mathematical derivations (in the asymptotic limit), we must not forget that they represent idealised notions, which have their ultimate foundation, in the last analysis, in reasonably successful but always problematic and only relatively certain predictions, experiments, and technical applications. A "formal rendition" is useful to clarify the logico-mathematical relations between different scientific theories, but we need to keep in mind that it has its ultimate foundations in nature as we know it through our practical-experimental interaction with the surrounding world.

7.5.2 Conclusion About Strong and Weak Intertheoretic Relations

Thus, we are pushed toward the following general conclusions: the distinction which has been drawn and made use of in our analysis, between weak and strong intertheoretic relations, is blurred. It must be abandoned in an important sense when it is recognized that both types of relation cannot exist in their pure form. They are only idealised types between which actual explanations and reductions must necessarily find a place. The distinction between two sorts of relations was provisional in a sense similar to that in which the falsework of a masonry arch may be removed when the arch is complete: though useful (or sometimes necessary) to obtain the final product, the falsework is not actually to be found in it, which nevertheless retains its equilibrium and its own specific reality; or better and less metaphorically, it is a distinction between two idealised and counterfactual limits, which we assumed in order to come to an indefinite number of intermediate forms of effective relations between different knowledge claims.

Finally, if the distinction between weak and strong intertheoretic reductions is abandoned, it is important to add that, from the perspectivalist point of view developed above, there is no reason why this conclusion should not be extended not only from the intertheoretical to the interlevel case (and therefore to the notions of mechanism, level, and component), but also from physics to biology, as I shall try to clarify now, in the last part of my paper.

7.6 Reductions and Multi-level Reality

As mentioned above, many exponents of the mechanistic philosophy besides Wimsatt have realized that interlevel explanations in terms of mechanisms and intertheoretic reductions are intimately connected. For this reason, among the traditional epistemological problems for the solution of which the concept of "mechanism" would supply the key, Machamer et al. 2000 mentioned intertheoretic reduction. Their strategy was to downplay intertheoretic reductions to the point of making them irrelevant for neuroscience and molecular biology. In these fields, theory change would be accurately described "in terms of the gradual and piecemeal construction, evaluation and revision of multi-level mechanism schemata" (Machamer et al. 2000: 23, quoted above).

However, this is fine if by this it is only meant that it is a mere *fact* that neuroscience and biology describe theory change "in terms of the gradual and piecemeal construction, evaluation and revision of multi-level mechanism schemata". But this seems also to be a shortcut to the conclusion that neuroscience and biology *should* operate this way. Then the question naturally arises: why are classical intertheoretic reductions, concerning more global relations between general or fundamental theories, in principle excluded from these fields of inquiry? Is there some principled reason to think so?

From the point of view of the mechanistic philosophy, I think that the answer should be a clear "no." But if this is right, why do mechanistic philosophers seem to incline towards a positive answer? One of the most important reasons is clearly suggested in the passage already quoted: they assume "logical deduction" and "the static two-place relations between different theories (or levels)" as essential ingredients of classical models of intertheoretic reduction.

However, as I have argued in the previous section, the distinction of weak and strong intertheoretic relations is blurred. As Hull already pointed out in a not too different context, strong reduction (in our terminology: the strongest intertheoretic relation) and replacement (in our terminology: the weakest intertheoretic relations) are the ends of a continuum of reductions (cf. Hull 1973: 633; Hull 1979: 317).

Thus, the most important basis of mechanistic philosophers for maintaining that classical intertheoretic reductions are in principle excluded from biology is removed. On the one hand the Nagelian model of intertheoretic reduction is only a regulative ideal (and as such it cannot ever be fully achieved, not just in contemporary neuroscience, but also in physics), and all reductions are "local" (Schaffner), or apply in a "dappled world" (to use Cartwright's term). But on the other hand, as a "regulative ideal", this model is an essential part of the dynamics of science, so that all our considerations, without exception, may be applied to physics, as well as to chemistry, biology, and any other empirical science. As far as the notion of progress is concerned, its elucidation, at least in very general terms, is relatively simple: in all cases (weak or strong), progress is an integration of different perspectives. Put differently, but again speaking in very general terms, each reduction is unification, however minimal.

From this point of view, weak kinds of intertheoretic reduction, based on weak analogies, are to be regarded, not only in neuroscience and biology, but also in physics, as perfectly legitimate types of reductions. The reductive part not only of Schaffner's model, but also of similar models (including, at least in some measure, some exponents of the already mentioned new wave model: see footnote 11, above) must be rejected and replaced by a much more flexible and adaptable kind of reduction, *which does not exclude any of the intertheoretic and interlevel reductions (now simply understood as ways of getting broader and more inclusive perspectives) that were de facto achieved by scientists in the course of the history of science and that experience, up to the present time, proved to be successful in some respects*.

Here it is important to emphasize that the impossibility of providing a Nagelian reduction that is both internally consistent and accurately represents what *de facto* goes on in the empirical sciences depends in the last analysis upon the fact that actual reductions are not to be decided a priori, but only by free discussion among scientists. These scientists are, not merely in principle, but also to a large extent in reality, led by a regulative ideal to pursue the unification of concepts, hypotheses, and scientific fields, between which up to their present moment of time no systematic (and, in the last analysis, experimental or practical-technical) relationships obtain.

Philosophers of science and methodologists may surely provide useful taxonomies for intertheoretic or interlevel reductions. Of course, there are many different ways of classifying such reductions, and the resulting taxonomy will differ as one point of view or the other is taken (for example their logical form, their use, their purposes, etc.). However, as Scheibe rightly maintains, the number of the possible reductions must be left somewhat open because we do not know what new kinds of reduction we will have to introduce in order to understand the further development of physics (Scheibe 1997, 2001: for example 353). In fact, all these taxonomies presuppose the existence of intertheoretic or interlevel relationships that have already been found by scientists in the course of their practice, in their effort to improve actual scientific theories by clarifying their relation to the past history of science, and especially to those earlier scientific theories that are in competition with them because of the fact that they provide clearly different solutions for similar problems.

From this point of view, relations of the kind specified by Schaffner in his first papers on this theme may be accepted at least in one important sense, while at the same time agreeing with Wimsatt that a non-formal (heuristic) account of reduction is not only allowed, but the only kind to be pursued.[12] We have seen that, according to Craver, (strong) reduction "is so peripheral to the practice of neuroscience that it is misleading to think of it as a regulative ideal for integrating neuroscience" (Craver 2007: 18, quoted above; but see Craver's Chapter 7 in general). On the contrary, according to him, the "mosaic unity" of neuroscience is sufficiently flexible to be able to cope with and interpret all kinds of intertheoretic relations. However, from what has been said, it will be clear that simply discarding the idea of a 'strong' kind of intertheoretic reduction, as Craver does by substituting it with his mosaic unity, is at best an oversimplification, and at worst a complete mischaracterization of the varieties of intertheoretic relations as set up by different theories and scientists. In particular, nothing in what has been said conflicts with the fact that, for instance, biology can, at least in principle, set up intertheoretic and interlevel reductions in a way that is analogous to the way in which physics does, as I shall now show, though very briefly, by an example taken from cancer research that in the last analysis seems to be less similar to the ideal limit of a weak intertheoretic or interlevel reduction (or to the model of Craver's "mosaic unity") than to the ideal limit of a strong reduction.

[12] Wimsatt 1974: 674. My references to this author should have made it clear to the reader that there are many similarities between the main theses I have presented here and Wimsatt's philosophy of biology. In order to avoid any misunderstanding, however, it is worth pointing out that there are also some differences, among which that concerning the notion of "perspective", already interesting enough on its own, is particularly important in our context. I started from a single general definition of "perspective", while Wimsatt distinguishes two meanings of the term. According to the first meaning – which is influenced by Uexküll's notion of "Umwelt" and for which he would now prefer the phrase "cross sections"– perspective is "most explicitly keyed to the point of view of a particular kind of organism or observer" (Wimsatt 2007: 230). According to the second meaning, perspective is ultimately defined in terms of its capacity to describe, "in a systematic way that is not level-bound", "trans-level interactions for such things as functional organization" (Wimsatt 2007: 212 and 230; see also the chapter 9); for the second meaning he gives as examples "[a]natomy, physiology, and genetics" (Wimsatt 2007: 231). It is easy to see that this distinction roughly parallels my distinction between, on the one hand, perspectives which may be used to identify levels, explanations, and mechanisms (and which stand in what I have called "weak relations" with one another), and, on the other hand, perspectives which have overlapping domains of application and are in competition with one another because they provide alternative explanations of the same phenomena (and which stand in what I have called "strong relations" with one another). There is, however, a fundamental difference. According to Wimsatt, the second meaning of "perspective" – which he had introduced at least since Wimsatt 1974 – is "not usefully captured by any of the notions of perspective discussed so far" (Wimsatt 2007: 230), that is, it is not included in the first. In other words, the second meaning differs in principle from that which, in Wimsatt's words, has to do with "a class of perspectives that map compositionally to one another so that their entities are related without cross-cutting overlaps in a hierarchical manner." (Wimsatt 2007: 229; italics in the original). The problem with this distinction is that it seems to exclude the possibility of a general meaning that serves as a common ground upon which Wimsatt's two meanings of "perspective" (and what I have called 'weak' and "strong" relations) can break a friendly lance with one another. In the first place, as I have urged before, this seems to be necessary to address and perhaps resolve the question of intertheoretic reduction. Second, and most importantly, by excluding this possibility, Wimsatt contradicts himself, challenging or undermining his claim – which I have taken up and developed as one of the fundamental theses of this paper – that we must be prepared to count reducibility as a matter of degree.

7.7 Cancer Research as a Possible Example of Interlevel/Intertheoretic Strong Reduction in Biology

The older 'paradigm' which attempts to explain how cancer arises and progresses is represented by the "Somatic Mutation Theory" (or SMT). In its early form, the theory was first formulated by Theodor Boveri, who argued that the main cause of the propensity for unrestrained proliferation is "a specific faulty assembly of chromosomes as a consequence of an event [*infolge eines Vorgangs*]." (cf. Boveri 1914/2008: 1; English transl. slightly modified). According to this approach, cancer is to be explained primarily at the cellular, subcellular and molecular levels of organization: cancer begins with alterations of *genes* in a single cell that are a necessary and sufficient condition to get cancer. Soto and Sonnenschein 2004 summarize the premises of SMT as follows: "(1) cancer is derived from a single somatic cell that has accumulated multiple DNA mutations, (2) the default state of cell proliferation in metazoa is quiescence, and (3) cancer is a disease of cell proliferation caused by mutations in genes that control proliferation and the cell cycle." (Soto and Sonnenschein 2004: 1097)

The rival 'paradigm' or, as Soto and Sonnenschein call it, the "alternative research program", is the "Tissue Organization Field Theory", according to which, on the contrary, a simple functional attribution to molecular parts does not fully account for the physiopathological phenomenon in question; carcinogenesis is rather to be explained as "a tissue-based disease", comparable to organogenesis, and proliferation is the default state of all cells. In this case, explanation comes at a different level, namely, at a level of tissue functional organization: carcinogens (directly) and mutations in the germ-line (indirectly) alter the normal interactions between the diverse components of an organ, such as the stroma and its adjacent epithelium. (For some of the more important scientific, epistemological and methodological aspects of this discussion, see above all Soto and Sonnenschein 2004, 2006, 2011, Soto et al (eds) 2016, and Bertolaso 2016, upon which my discussion is mainly based and where further references will be found).

The two theories have been set up in direct opposition to one another and championed respectively by Vaux 2011 and Soto and Sonnenschein 2011. On the one hand, the somatic mutation theory is supported by observations of leukemias that bear specific chromosome translocations (Vaux cites as an example Burkitt's lymphoma and chronic myeloid leukemia). Although the SMT had to be modified because of some *prima facie* evidence to the contrary (for example, epigenetic inheritance, tumour progression through accumulation of further mutations), it is validated by the successful treatment of certain malignancies with drugs that directly target the product of the mutant gene (Vaux 2011). On the other hand, it is maintained that the structure of SMT is now similar to that of pre-Copernican astronomy, with new epicycles added when something does not fit. Some types of cancer are not associated with any mutations whatsoever. Data suggest "the coexistence of apparently normal organs at the morphological level of organization (anatomical, histological and cellular architecture) with significantly altered gene expression."

(Soto and Sonnenschein 2006: 365–355). It is not surprising, then, that somatic mutations are questioned as representing "the" cause for the majority of cancers (Brücher and Jamall 2016: 1664).

In the light of the information presently available, it seems to me that Craver's "mosaic unity" will not be a very probable result, at least in the sense that, as he says in a passage already quoted, the mosaic unity of neuroscience is given by different converging lines of research carried out by "autonomous" (and therefore 'compatible') fields, with different central problems, different techniques, different theoretical vocabularies, and different background assumptions (cf. Craver 2007: 231). To the contrary, although the Somatic Mutation Theory and the Tissue organization Field Theory give explanations at different levels of organization of a biological system, they are clearly incompatible explanations of the same phenomenon: they are, as Soto and Sonnenschein say, "alternative" theories of carcinogenesis (Soto and Sonnenschein 2011: 332). For example, the two perspectives on cancerogenesis are incompatible because they are founded upon contradictory assumptions concerning the default state of cells: according to SMT, the default state of cells is quiescence, whereas for the Tissue organization Field Theory cells display 'agency' and proliferate unless constrained. Or again, one theory maintains that a cell can influence a tissue (undermining its functionality), whereas the other insists that tissues can influence cells both negatively and positively (as when cancer cells develop into healthy cells if implanted into a healthy tissue).

Surely, we do not know, and it would be hazardous to suppose, whether the one, the other, or perhaps a third new theory will prevail. SMT is perhaps only wrong for most, but not all cancers, as for example Brücher and Jamall 2016 have maintained. But it is interesting to notice that the very possibility of the third case entails the possibility of strong intertheoretic and interlevel relations in biology. For in this case a new theory would prevail by providing an integrative explanation and introducing new mechanisms, level or components. In other words, this case would offer a good representative of the 'strong' intertheoretic relationships with which we have been dealing, which occur much more rarely in biology than physics, but which are not a priori or in principle excluded from any scientific field whatsoever.

7.8 Conclusion

A satisfactory account of interlevel explanation requires a perspectival conception according to which different mechanisms, levels, and components correspond to different perspectives, which in turn depend upon different pragmatic interests and practical possibilities of the inquirers concerned (Sect. 7.2). However, a perspectival conception is only a necessary step towards a satisfactory account of interlevel explanation, since it cannot explain how interlevel explanations may interact or even clash with one another when trying to solve the same problem. Provided that different mechanisms, levels, and components are defined by different theoretical perspectives, an acceptable solution to the question of *interlevel* explanations - as

Wimsatt (1974, 2006) had already supposed – must go hand in hand with a tenable solution of the old and thorny problem concerning *intertheoretic* reduction.

This problem was briefly dealt with in the second part of the paper. For this purpose, I distinguished in Sect. 7.3, though only provisionally, between two kinds of relation concerning theoretical perspectives: a *weak* interlevel/intertheoretic relation, which holds when explanations offered from different and autonomous perspectives contribute to a better understanding of the same phenomena; and a *strong* interlevel/intertheoretic relation, in which theories are in competition with one another. Sections 7.4 and 7.5 proffered respective analyses of the 'weak' and the 'strong' kinds of relation between the theories/levels that scientists work with. Finally, this distinction had to be removed because both types of relation are only ideal and counterfactual limits, never reached in practice. In Sect. 7.6 I pointed out some other important similarities between weak and strong relations: both have in common that relating multiple perspectives to one another in order to better understand the subject-matter under investigation requires constructing a new, wider or deeper perspective; and in both cases, the question of interlevel explanatory reductions, just as that of intertheoretic reductions, cannot be answered by purely abstract philosophical considerations, but only with reference to, and in accordance with, the practice of scientists and the history of science. This is true in principle not only for physical, but also for all empirical theories, as I tried to show, though briefly, in Sect. 7.7 by presenting an example taken from current cancer research.

From the foregoing considerations we may draw a last and very general methodological lesson, which is not to be confused with the kind of continuity hoped for by the logical empiricists: researchers always have to be not only selective and prepared to use several, multilevel perspectival accounts to inform one another in connection with specific problems, but also to be integrative, in the sense of trying to establish interrelations — sometimes weaker, sometimes stronger — holding between different theoretical perspectives and, at the same time, between the ontological mechanisms, levels and components that these perspectives express.

Acknowledgments I presented an earlier and briefer version of this paper at the Congress of the "Académie International de Philosophie des Sciences" at the University of Dortmund, Germany (October 28–30, 2016). I am grateful to Stuart Glennan, Brigitte Falkenburg, Georg Schiemann and Mike Stuart for constructive criticism and helpful comments on an earlier draft of this article.

References

Agazzi, E. 2014. *Scientific Objectivity and Its Contexts*. Berlin: Springer.

Batterman, R.W. 2002. *The Devil in the Details. Asymptotic Reasoning in Explanation*. Oxford: Oxford University Press.

Bechtel, W. 2006. *Discovering Cell Mechanisms: The Creation of Modern Cell Biology*. Cambridge: Cambridge University Press.

———. 2008. *Mental Mechanisms. Philosophical Perspectives on Cognitive Neuroscience*. London: Routledge.

Bertolaso, M. 2016. *Philosophy of Cancer. A Dynamic and Relational View*. Dordrecht: Springer.

Bertolaso, M., and M. Buzzoni. 2017. Causality and Levels of Explanation in Biology. In *Philosophical and Scientific Perspectives on Downward Causation*, ed. M. Paolini Paoletti and F. Orilia, 164–179. London: Routledge.

Bickle, J. 1998. *Psychoneural Reduction: The New Wave*. Cambridge, MA: MIT Press.

———. 2003. *Philosophy and Neuroscience: A Ruthlessly Reductive Account*. Boston: Kluwer.

Bitbol, M. 2014. Downward Causation Without Foundations. In *The Philosophy of Chemistry: Practices, Methodologies, and Concepts*, ed. Jean-Pierre Llored, 675–705. Newcastle: Cambridge Scholar Publishing.

Boveri, T. 1914/2008. *Zur frage der entstehung maligner tumoren*. Jena, Germany: Gustav Fisher. English Transl. by Henry Harris. 2008. Concerning the Origin of Malignant Tumours. *Journal of Cellular Science* 121(Suppl 1): 1–84.

Brücher, B., and I.S. Jamall. 2016. Somatic Mutation Theory – Why It's Wrong for Most Cancers. *Cellular Physiology and Biochemistry* 38: 1663–1680.

Buzzoni, M. 1995. *Scienza e tecnica. Teoria ed esperienza nelle scienze della natura*. Rome: Studium.

———. 1997. Erkenntnistheoretische und ontologische Probleme der theoretischen Begriffe. *Journal for General Philosophy of Science* 28: 19–53.

———. 2008. *Thought Experiment in the Natural Sciences*. Würzburg: Königshausen+Neumann.

———. 2016. Mechanisms, Experiments, and Theory-Ladenness: A Realist–Perspectivalist View. *Axiomathes* 26: 411–427.

———. forthcoming. Thought Experiments in Philosophy: A Neo-Kantian and Experimentalist Point of View. *Topoi* in press (online 2016). https://doi.org/10.1007/s11245-016-9436-6.

Chen, Y., F. Ding, H. Nie, A.W. Serohijos, S. Sharma, K.C. Wilcox, S. Yin, and N.V. Dokholyan. 2008. Protein Folding: Then and Now. *Archives of Biochemistry and Biophysics* 469: 4–19.

Churchland, P.M. 1979. *Scientific Realism and the Plasticity of Mind*. Cambridge: Cambridge University Press.

———. 1984. *Matter and Consciousness: A Contemporary Introduction to the Philosophy of Mind*. Cambridge, MA: MIT Press.

Craver, C.F. 2001. Role Functions, Mechanisms, and Hierarchy. *Philosophy of Science* 68: 53–74.

———. 2007. *Explaining the Brain. Mechanisms and the Mosaic Unity of Neuroscience*. New York: Oxford University Press.

———. 2013. Functions and Mechanisms: A Perspectivalist View. In *Functions: Selection and Mechanisms*, ed. Philippe Huneman, 133–158. Berlin: Springer.

Craver, C.F., and J. Tabery. 2017. Mechanisms in Science. In *The Stanford Encyclopedia of Philosophy* (Spring 2017 Edition), ed. Edward N. Zalta. https://plato.stanford.edu/archives/spr2017/entries/science-mechanisms/.

Darden, L. 2008. Thinking Again About Biological Mechanisms. *Philosophy of Science* 75: 958–969.

De Regt, H., and D. Dieks. 2005. A Contextual Approach to Scientific Understanding. *Synthese* 144: 137–170.

Dizadji-Bahmani, F., R. Frigg, and S. Hartmann. 2010. Who's Afraid of Nagelian Reduction? *Erkenntnis* 73: 393–412.

Duhem, P. 1914. *La Théorie physique. Son objet et sa structure,* 2nd ed. Paris: Chevalier & Riviere.

Endicott, R. 1998. Collapse of the New Wave. *Journal of Philosophy* 95: 53–72.

———. 2001. Post-Structuralist Angst—Critical Notice: John Bickle, Psychoneural Reduction. *Philosophy of Science* 68: 377–393.

Eronen, M.I. 2013. No Levels, No Problems: Downward Causation in Neuroscience. *Philosophy of Science* 80: 1042–1052.

———. 2015. Levels of Organization: A Deflationary Account. *Biology and Philosophy* 30: 39–58.

Falkenburg, B. 2007. *Particle Metaphysics. A Critical Account of Subatomic Reality*. Berlin/Heidelberg: Springer.

Feyerabend, P.K. 1966. The Structure of Science. *The British Journal for the Philosophy of Science* 17: 237–249.
———. 1993. *Against Method*. London: Verso (3th edition)
Franklin, A. 1989. The Epistemology of Experiment. In *The Uses*, ed. D. Gooding, T.J. Pinch, and S. Schaffer, 437–460. Cambridge: Cambridge University Press.
Glennan, S. 2002. Rethinking Mechanistic Explanation. *Philosophy of Science* 69: S342–S353.
Goosens, W.K. 1978. Reduction by Molecular Genetics. *Philosophy of Science* 45: 73–95.
Gutschmidt, R. 2014. Reduction and the Neighbourhood of Theories. *Journal for General Philosophy of Science* 45: 49–70.
Heisenberg, W. 1959. Die Plancksche Entdeckung und die philosophischen Problerne der Atomphysik. *Universitas* 14: 135–148.
Hooker, C.A. 1981. Towards a General Theory of Reduction. Part III: Cross-Categorial Reductions. Dialogue 20 (Part I: Historical and scientific setting, 38–59. Part II: Identity in reduction, 201–236. Part III: Cross-categorial reduction, 496–529).
———. 2004. Asymptotics, Reduction and Emergence. *The British Journal for the Philosophy of Science* 55: 435–479.
Hull, D.L. 1973. Reduction in Genetics – Doing the Impossible. In *Logic, Methodology and Philosophy of Science IV*, ed. P. Suppes et al., 619–635. Amsterdam: North-Holland.
———. 1979. Reduction in Genetics. *Philosophy of Science* 46: 316–320.
Kemeny, J., and P. Oppenheim. 1956. On Reduction. *Philosophical Studies* 7: 6–19.
Kimbrough, S.O. 1979. On the Reduction of Genetics to Molecular Biology. *Philosophy of Science* 46: 389–406.
Klein, C. 2009. Reduction Without Reductionism: A Defence of Nagel on Connectability. *Philosophical Quarterly* 59 (234): 39–53.
Krüger, L. 1974. Wissenschaftliche Revolutionen und Kontinuität der Erfahrung. *Neue Hefte für Philosophie, Heft* 6-7: 1–26.
Kuhn, T.S. 1970. *The Structure of Scientific Revolutions*. 2nd ed. Chicago: University of Chicago Press.
Machamer, P., L. Darden, and C.F. Craver. 2000. Thinking About Mechanisms. *Philosophy of Science* 67: 1–25.
Nagel, E. 1949. The Meaning of Reduction in the Natural Sciences. In *Science and Civilization*, ed. R.C. Stauffer, 99–135. Madison: University of Wisconsin Press.
———. 1961. *The Structure of Science*. London/New York: Harcourt-Brace.
Needham, P. 2010. Nagel's analysis of reduction: Comments in defence as well as critique. *Studies in History and Philosophy of Modern Physics* 41: 163–170.
Nickles, T. 1973. Two Concepts of Intertheoretic Reduction. *The Journal of Philosophy* 70: 181–201.
———. 1988. Reconstructing Science: Discovery and Experiment. In *Theory and Experiment: Recent Insights and New Perspectives on Their Relation*, ed. D. Batens and J.P. van Bendegem, 299–333. Dordrecht: Reidel.
O'Malley, M.A., I. Brigandt, A.C. Love, J.W. Crawford, J.A. Gilbert, R. Knight, S.D. Mitchell, and F. Rohwer. 2014. Multilevel Research Strategies and Biological Systems. *Philosophy of Science* 81: 811–828.
Pâslaru, V. 2009. Ecological Explanation Between Manipulation and Mechanism Description. *Philosophy of Science* 76: 821–837.
Popper, K.R. 1972. *Objective Knowledge: An Evolutionary Approach*. Oxford: Clarendon Press.
Potochnik, A., and B. McGill. 2011. The Limitations of Hierarchical Organization. *Philosophy of Science* 79: 120–140.
Primas, H. 1985. Kann Chemie auf Physik reduziert werden? Chemie in unserer Zeit 19: 109–119 (Erster Teil: Das Molekulare Programm), 161–166 (Zweiter Teil: Die Chemie der Makrowelt)
Rueger, A., and P. McGivem. 2010. Hierarchies and Levels of Reality. *Synthese* 176: 379–397.
Schaffner, K.F. 1967. Approach to Reduction. *Philosophy of Science* 34: 137–147.
———. 1977. Reduction, Reductionism, Values, and Progress in the Biomedical Sciences. In *Logic, Laws, and Life*, ed. R. Colodny, 143–171. Pittsburgh: University of Pittsburgh Press.

———. 1992. Philosophy of Medicine. In *Introduction to the Philosophy of Science*, ed. M.H. Salmon et al., 310–345. Cambridge: Hackett.

Schaffner, K. 2012. Ernest Nagel and Reduction. *The Journal of Philosophy* 109: 534–565.

Scheibe, E. 1973[2001]. Die Erkärung der Keplerschen Gesetze durch Newtons Gravitationsgesetz. In: Einheit und Vielheit. Festschrift für CF von Weizsäcker zum 60. Geburtstag. Ed. by E Scheibe and G Süssmann, Vandenheock & Ruprecht: Göttingen, pp. 98–118. English transl.: The Explanation of Kepler's Laws, in: Scheibe (2001): 306–323.

———. 1993[2001]. Heisenbergs Begriff der abgeschlossenen Theorie. In: Werner Heisenberg. Physiker und Philosoph. Ed. by B.Geyer et al., p. 251–257. Spektrum: Heidelberg. English transl.: Heisenberg's Conception of a Closed Theory. In: Scheibe (2001): 136–141 (quotations are from the English edition).

———. 1997. *Die Reduktion physikalischer Theorien. Ein Beitrag zur Einheit der Physik. Teil I: Grundlagen und elementare Theorie*. Berlin: Springer.

———. 1999. *Die Reduktion physikalischer Theorien. Ein Beitrag zur Einheit der Physik. Teil II: Inkommensurabilität und Grenzfallreduktion*. Berlin: Springer.

———. 2001. *Between Rationalism and Empiricism*, Selected Papers in the Philosophy of Physics, ed. B. Falkenburg. Berlin: Springer

Soto, A.M., and C. Sonnenschein. 2004. The Somatic Mutation Theory of Cancer: Growing Problems with the Paradigm? *BioEssays* 26: 1097–1110.

———. 2006. Emergentism by Default: A View from the Bench. *Synthese*. 151: 361–376.

———. 2011. A Testable Replacement for the Somatic Mutation Theory. *Bioessays* 33: 332–340.

Soto, A.M., G. Longo, and D. Noble, eds. 2016. From the Century of the Genome to the Century of the Organism: New Theoretical Approaches. *Progress in Biophysics and Molecular Biology* 122 (1): 1–82.

Strand, A., and G. Oftedal. 2009. Functional Stability and Systems Level Causation. *Philosophy of Science* 76: 809–820.

Stuart, M.T. 2016. Taming Theory with Thought Experiments: Understanding and Scientific Progress. *Studies in History and Philosophy of Science* 58: 24–33.

———. 2018. How Thought Experiments Increase Understanding. In *The Routledge Companion to Thought Experiments*, ed. M. Stuart, Y. Fehige, and J.R. Brown, 526–544. New York: Routledge.

Vaux, D.L. 2011. In Defense of the Somatic Mutation Theory of Cancer. *BioEssays* 33: 341–343.

von Weizsacker, C.F. 1971. *Die Einheit der Natur*. München: Weizsacker C F von. Die Einheit der Natur. München.

Weber, M. 1922[1949]. Gesammelte Aufsätze zur Wissenschaftslehre, Tübingen: Mohr. Translated and Edited by Edward A. Shils and Henry A. Finch, On The Methodology of the Social Sciences, pp. 113–188. Glencoe (Ill.): Free Press (quotations are from this edition).

Wimsatt, W.C. 1974. Reductive Explanation: A Functional Account. *PSA: Proceedings of the Biennial Meeting of the Philosophy of Science Association*: 671–710.

———. 2006. Aggregate, Composed, and Evolved Systems: Reductionistic Heuristics as Means to More Holistic Theories. *Biology and Philosophy* 21: 667–702.

———. 2007. *Re-Engineering Philosophy for Limited Beings: Piecewise Approximations to Reality*. Cambridge: Harvard University Press.

Winther, R.G. 2009. Schaffner's Model of Theory Reduction: Critique and Reconstruction. *Philosophy of Science* 76: 119–142.

Chapter 8
A Methodological Interpretation of Mechanistic Explanations

Hans Lenk

Abstract Some current approaches to causal-mechanistic and/or deterministic analyses in epistemology are to be discussed from the vantage point of the author's methodological scheme-interpretationism. The assets of the new mechanism orientation are, next to an "indirect" realist interpretation of the differentiated (old and "new") mechanisms, primarily a study of the "producing" intrinsic processes of causal "mechanisms". The analysis of effect-engendering/productive and relatively stable or repeatable continuous (chain or branching) processes – i.e. "new" mechanisms – are well suited to describe and answer the very intriguing internally differentiated "how come" and why-questions.

8.1 Introduction: Methodological Scheme-Interpretationism as an Epistemological Approach

We have to be able to compare, describe, structure, explain, conceptualize, understand, identify and 'grasp' by somehow deciphering 'existing' relations – i.e. by some means of interpretation. We cannot do without using patterns, forms, schemes, models, networks, configurations and representations of structures: In our cognitions of any kind we are obliged to use such forms, shapes, frames and constructs – all of them are *schemes*. I use that concept as a comprehensive term to cover all these 'structures' and their activation and representation. This applies to all sorts of 'grasping' something, may it be by a process of cognition and categorization or of normative structuring or by planned acting (see my 2003). Applications of forms and frames are "schematizations", whether performed consciously or subconsciously. I would like to call these "interpre(ta)tive" or "*interpretation constructs*" (see my 1993a, b, passim) and their activation "*scheme*-interpretations" in order to distinguish them from the usual text interpretation in the hermeneutical sense.

H. Lenk (✉)
Karlsruher Institut für Technologie, Institut für Philosophie, Karlsruhe, Germany
e-mail: hans.lenk@kit.edu

B. Falkenburg, G. Schiemann (eds.), *Mechanistic Explanations in Physics and Beyond*, European Studies in Philosophy of Science 11,
https://doi.org/10.1007/978-3-030-10707-9_8

Indeed, we always have to "*scheme-interpret*", as I say, i. e. to use and activate schemes and any kinds of the mentioned structural forms and representations: we simply cannot *not* interpret (see my 1993a, b, 350; 2003), we *must* always scheme-interpret in any cognition, action and in all kinds of modeling and representing whether linguistic or not.

Any recognizing and generalizing, particular conceptual knowledge is bound to cognitive or action-guiding (sometimes even normative) schemata. These can be understood as more or less abstract constructs which are projected onto and into the seemingly direct sense perception and the respective experiences by recognizing *Gestalten* or constituting objects, processes, events etc. Any seeing and recognizing shapes and forms is dependent on and guided by schemes.[1] Any cognition whatsoever is thus schematic or scheme-laden or -imbued etc. Therefore, I speak of "*scheme--interpretation*". Methodologically speaking, that sort of *(scheme-)interpretation is* but the *activation and reactivation of schemes*. Scheme-Interpretation according to my terminology is indeed the development, stabilization and *activation* (application) *of* mentally representing constructs or *schemes*. *Interpretation* (in a wide sense though) *is basically scheme-interpretation* and founded on as well as grounded in scheme activation. We can even conceive of a basic axiom or *principle of methodological (scheme-)interpretationism* stating that all kinds of grasping, cognition and action are interpretation-dependent, i.e., founded on the activation of schemes. This is true far beyond psychological theories and epistemological perspectives, but rather a totally general methodological comprehensive approach comprising the philosophy of knowledge (traditionally called epistemology) as well as a philosophy of (structuring) action and representation. Any kind of interpretation whatsoever is connected with or bound to an activation of such schemes. This connection might be characterized by core features and essential stimuli the very selection of which is necessary, even though many of these are conducted subconsciously.[2] As mentioned schemes might be used consciously or often activated subconsciously.

*The approach is quasi Kantian, yet more flexible (*see below) or even *a Cassirian symbolic construct-* or *scheme-interpretationism: It would somehow overarch the splits between natural and the social sciences as well the humanities* (see my 1993a, b, chaps. 17 f; 2006, 26–39; 2007, 35–47; 2017, chap. 3). All these disci-

[1] It is important to notice that schemes also consist of sub-schemes. The activation of a sub-scheme is usually immediately related with the activation of the schema itself and the other way around. The comparison of schemata with programs, networks etc. is certainly fruitful and can be visualized in flow charts [even networks] and related structural means admitting of state and point identification of the constituents and the ramifications of some such structures. The encompassing set of the schemes we use to interpret "our world" would represent and comprise in a sense our "private theory" (Rumelhart 1978) of the nature of [our] "reality". Schemes represent or mirror so to speak our internal models of the respective situations in the world. For cognitions and actions they must be activated, i.e., occur in/as interpretation (processes).

[2] On the subconscious level, cognitive quasi-constructs are used to render the profiles of contrast and the structural differentiation by activating the functions of the respective sense organs or their processing units of perception and cognition in the brain as well as the integrating poly-modal and combining (yet in part hypothetical or even hypostatized) centres.

plines would structure/order their fields and objects according to the activation of schemata by using procedures of establishing, stabilizing and activating schemes as cognitive constructs in order to structure the respective field of subject matters. It is important to note that this is true also for all-today cognitions and actions.

Interestingly enough, scheme-interpretation admits of levels of differentiation according to the variability of the respective schemes or meta-schemes and their respective activations. An example would be the question, whether or not they are hereditarily fixed or conventionalized or flexible. They are partially due to hereditary and evolutionary development, or developed by early ontogenetic interaction with 'the world', whether learned by experience, design or instruction. I developed a hierarchy of levels of interpretation consisting of six different levels,[3] $\mathbf{IL_1}$ through $\mathbf{IL_6}$.

Thus, these schemes are developed and applied on different representational levels in order to integrate individual experiences, single activities and sense data or stimulations into a more comprehensive or general frame, pattern or similarity structure.[4] Whenever we try to identify, retrieve, recognize shapes transcending the

---

[3] The different levels of interpretation are the following ones: IL1 comprises the practically unchangeable productive primary interpretations of primary constitution which might be represented by subconscious schema instantiation. They comprise the hereditarily fixed or could have these senses or at least devise technological means for substituting these. – On the second level we have the habitual, quality forming frame interpretations and schema categorisations as well as 'categorizations' that are abstracted from pre-linguistic discriminatory activities, experiences of equality of shape, similarity of presentation and experience etc. Establishment and discriminatory capacity of pre-linguistic conceptualization and development of concepts about language are to be formed on this level. – On level IL_3 we have conventional concept formation, namely socially and cultural traditional conventions and norms for representation and forms of discriminatory activities like the explicit conceptualization of framing the world according to natural kinds etc. In so far as this is not related already to language differentiation we can think of a sublevel (IL_{3a}) on which pre-linguistic convention(alization)s are characteristic. On the other hand (on IL_{3b}) we have the explicitly linguistic conventionalization or the differentiation of concepts by means of language. – Level 4 would comprise the consciously formed interpretations of embedding and subsuming as well as classifying and describing according to generic terms, kinds etc. It is the level of ordered concept formation and classification as well as ordering and subsumption. – Level IL_5 would go beyond that by rendering explanatory, or in the narrower sense comprehending ("*Verstehen*") interpretations as well as justifying a theoretically argumentative interpretations in a sense of looking for reasons and grounds of justification. – Beyond that, it is theoretically important that we have also a level ($\mathbf{IL_6}$) of the epistemological and philosophical as well as methodological interpretations of a *meta*-character, overarching and integrating the procedures of theory building and theory interpretation, methodology and the models of interpretation in the sense of methodological scheme-interpretationism itself. One could call this a cumulative *meta*-level of interpretation and explicitly speak of epistemological *meta*-interpretations.

[4] In any case, whenever we try to combine phenomena and the results of categorizing under generic perspectives, concepts, equalities of form or shape and similarities as well as analogues genetically founded activation of selective schemata of sense perception (e. g. contrasts of dark and light etc.) as well as the interactive, selective activations of early ontogenetic developments like the stages of developmental psychology discussed by Piaget. – Also comprised are the biologically hardwired primary theories which we cannot alter at will, but which we can (only) problematize in principle. For instance we have no magnetic sense or capacity to trace ultrasound like the bats. But we can conceive of conditions in which we would have (*analogues*) of all these.

phenomenality of the so-called qualitatively *given*, we rely on the activation of such schemes. This is true certainly *a fortiori* for mechanistic approaches (see below.). – Thus far the necessary introductory remarks.[5]

8.2 Causal Interpretations. Describing and Explaining Causes and Causation by Scheme-Interpretation

From one perspective, causality is but made up or does consist of "causal relationships", e.g., relations between traditionally speaking of (rather complex constellations of) "causes" on the one hand and "effects" on the other (see Beebee et al. 2012). We can thus speak of 'causal interpretations' or "causal explanations" of "causal relations", events, phenomena etc., if we connect specific (descriptions of) events or states of affairs with later events or states in a regular manner, if not as a practically 'necessary' sequence. Or if the occurrence or ongoing of the first such ones is always followed by an obviously regular effect or impact without exception, i.e., if not disturbed from 'outside'. The earlier events, states etc. will be followed in (a mostly chain-like) time-succession by the latter ones, if no other disturbing event or circumstance interferes. (But that quasi-'necessity' is as a rule *not a logical or a priori 'necessary'* sequence, consequence or conclusion.)

On the other hand, causality is also not just an empirical sequence either, but (can be) – pragmatically speaking – a factually exception-less sequence. At least, it should be interpreted that way by switching to another layer of using a sort of semantically or theoretically higher-level or meta-level: it is indeed in the end amounting to a *methodological-epistemological turn* avoiding an all too strong ontological stance or presupposition.

We may, according to methodological-epistemological scheme-interpretationism (see above and my 2000/2007, 2003), cope with the problem situation and philosophical difficulties of an ontological provenance in the following way. We may take a *quasi-Kantian* epistemological or, rather, *methodological*, approach like the one sketched above. Thus, we would, e. g., address the deep-rooted reality "as such" or "the thing in itself" (Kant 1956[1781/1787] ff., A235 ff.|B294) – and the realism problem with notable extensions for the problem – by allotting indirectly to it a 'realistic' character. This amounts to taking an *epistemological-methodological turn*.

We may say that what we call 'the external world' as such or "thing in itself" should be interpreted as to 'have' such a sort of *quasi 'constitutive' quality or quasi-*

That is true not only for cognition, but also for actions, i.e. not only for passive sorts of 'grasping', but also for rather active kinds of cognition and action – even for normative (prescriptive or evaluative) patterns. – In general, we use mental representations of frames, structures, models and any other data features or contents which are typified, generically distinguished and concentrated to relevant features which are retrievable from memory.

[5]In English, see my (2000, 2003, chap. 1, 2, 2007, chaps. 2, 5). German originals (1993a, b, 1995a, b and recently 2017).

structure (basic constitution) that the methodology of scheme-interpretation as outlined above *of causal relation(ship)s/connections can be successfully carried through.* Thereby we would not hypostatize "direct" or any substantive mapping knowledge of the 'world as such' or "in itself" (in Kant's wording). We can and even should or must interpret the situation of world knowledge and interaction with "reality" by stating that the 'underlying' world would 'have' (i.e. could/should be indirectly assigned) the epistemological characteristic or quasi-'property' that practically so-called "causal relation(ship)s" (or, rather, such interpretations) can be successfully applied to the usually so-called "causal" connections and phenomena. (Here, it is not even logically or "a priori" "necessary" to hypostatize a Kantian "thing in itself" in the narrower sense as a "real" world structure.) In any case, Kant's very approach can be understood as just vindicating what he calls "*Betrachtungsweise*" (interpretative perspective). Methodolologically speaking it is a higher-level interpretation of a "version or mood of cognition" (*Erkenntnisweise*) by which we "interpret" (*deuten*) the "objects of cognition" (*Erkenntnisgegenstände*) instead of cognizing them direct(istical)ly as objects or "things in themselves" (see my 2007, chap. 5).

For a modified quasi-Kantian, yet essentially methodological, perspective, this would also apply to the interpretative perspective regarding the "*category*" of causality, i. e. causal interpretations and explanations, though in a rather 'liberalized'[6] understanding of a methodological-epistemological modification of a quasi-Kantian approach as mentioned. I am really tempted to read Kant as being a defender (or forerunner of) methodological interpretationism, in a sense at least (see my 2006).

Modern ramifications of causal perspectives and variants from classical determinism and factually nomological or "material causation" through action-oriented ("interventionist") and probabilistic, quantum theoretical and (non-consequentialist) "inclusive causation" amount to a rather wide "variety of causalities" (Sosa 1993, 240 f, Beebee et al. 2012). Only some of these that deal with the interpretation-realistic approach (see above and my 2003) can be discussed here only in short selections and by appreciative methodological remarks.

Brigitte Falkenburg (2012) also starts from Kant's compromise between the "too strong" rationalistic understanding of causality and the "too weak empiricist" one (ibid. 272–273). Yet, for Kant, the temporal order of cause and effect would not only

---

[6] Instead of Kant's "*a priori*" foundation of his "categories" as the necessary and universal even "logical" absolute structures of the "Understanding" ("*Verstand*" in his understanding) we have to give up Kant's absolutism (universal validity founded by the forms of Reason (*Vernunft*) and necessity logically derived from the logical forms of judgments, see my 1968, chap. I) This certainly is true also for a more precise and differentiated interpretation of (the concept/category of) *causality*. In this case as in general, we can use some of the Kantian categories in a more 'liberalized' fashion – allowing, e.g., statistical and even collective or chance causation like in thermodynamics or in the measurement problem of quantum mechanics.

Notwithstanding Kant's deterministic, even 'logical' absolutism he had an important inkling – via his epistemological and metaphysical 'indirectism' – i.e. a rather methodological interpretationism of sorts: he may be taken as a progenitor (or at least forerunner) of a more flexible category- or scheme-interpretationism.

be 'constitutive' but also "necessary" according to the "Principle and Category of Causality" stemming from Reason plus Understanding themselves and being derived from the "forms of judgment" (*Urteilsformen)* after Reich's reconstruction of Kant's deduction of the respective categories (see, e.g., my 1968, chap. I). Falkenburg rightly concludes that Kant's universally valid Principle of Causality essentially amounts to "a *methodological requirement*", namely the basic principle (*Grundsatz*) "for any effect *to ask* for the cause" (my ital.). Without assuming that methodological principle of such necessary connection we would have "no continuous experience" at all.

Passing through the famous causality interpretations from Mill[7] through Mackie[8] Falkenburg (ibid. 275) thinks that Kant's principle of causality, though often seen as outdated, is indeed "a jewel underestimated in its value" that should be re-established, though as a "*methodological*" strategy or "principle". In addition to Hume's empirical theory of regularity, an "*interventionist*" concept of causality would nicely "serve to understand experiments". Indeed, though it should not be seen as the one and .only offer in the store of relevant approaches. Falkenburg emphasizes:

> "Offers from neighboring stores in *physics*" are being "taken on philosophical sale or return", so to speak (ibid.).

All of these models do have their methodological and practical problems though, each with pros and cons regarding the respective field and its embedding in society and different sciences, including the social and biological sciences and practical applications like experimenting in, e.g., engineering etc.[9]

---

[7] Although J. St. Mill was an overall adherent of Hume's regularity theory of the so-called "causation", he wistfully wrote (in his 1843, 445): *"All laws of causation, in consequence of their liability to be counteracted, require to be stated in words affirmative of tendencies only, and not of actual reality"(my ital.)* .– That statement seems to me (and some other authors like Cartwright 1989; Wachter 2009; Falkenburg 2012) a very far-reaching insight from a methodological and epistemological point of view.

[8] Mackie's (1965/1993, 34 f, 1974, 62) famous INUS conditions ("*i*nsufficient but *n*ecessary part of a (complex, HL) condition which is itself *u*nnecessary but *s*ufficient for the result" (suggested by D.C. Stove) can indeed also be interpreted as a variant of a (potential) regularity conception of causation. That however depends on the rather open, if not unlimited disjunction of the "minimal sufficient conditions" (term after Marc-Wogau 1962) and the understanding of the sort of "necessitation" involved. The latter one need not be strictly "lawlike" or of DN-form but may be "parasitic on contingent general principles" (Sosa 1993, 241) or a "*set* of properties" being "a minimal sufficient set" or disjunction of such ones for the caused individual event/result (Kim 1993, 72). There are different non-lawlike "cases of necessitation each with its own distinguishing features" which we still (in particular in historical or counterfactual or probabilistic relations) may call some kinds of factually quasi-necessary variants of causation (Sosa ibid., 242).

[9] Other problems are raised by functional causal or teleonomic explanations as well as from statistical causality in quantum measurement problems or collective mass interaction explanation – like in thermodynamics quantum theory – or deterministic (or may be in the near future also probababilistic) chaos theory or in mathematically complex dynamics with bifurcation dynamics and strange attractors etc.

As regards physics, we indeed have a sort of "supermarket mixture" ("*ein physikalischer Gemischtwarenladen*", ibid. 282) of approaches even in different branches of physics, depending on the theoretical context, in lack of a unique fundamental physical theory (ibid., 280–282). Even in physics, and all the more in the general world of the sciences, we see that '*the*' alleged concept of causality did disseminate or disintegrate into "*a plurality of concepts*". It is indeed unclear how again to recombine or reintegrate this plurality (see the notorious problems of causality-related problems to integrate and explain 'indeterministically' described phenomena like radioactive decay on the one hand and so to speak "classical" relativistic light propagation description etc. on the other). Therefore, Falkenburg concludes that generally speaking "causality is a *pre-scientific concept splitting up into a plurality of concepts*" (ibid. 282). Instead of an unambiguous unified concept of causality, we have "the reconstruction of physical, electro-chemical, biochemical" so-called "*mechanisms*" (ibid. 288, see below) or, rather, *systems mechanisms, as also such ones in molecular biology and chemistry, neurophysiology etc.* We have thus to say goodbye to "mono-causal thinking" and give way to manifold ways of "*causal modeling*" taking into account "complex ensembles of conditions" and circumstances as well as environments and complex systems and systems interactions (ibid. 286).

Falkenburg's and my own approach are very similar – as regards the methodological accent. Primarily the 'liberated' quasi-Kantian solution, in between of 'strong' (necessary in the strictly factual or even sort of 'logical' sense) on the one hand and the 'weaker' regularity or even probabilistic etc. causalities on the other. It is to my mind indeed a viable and feasible way out of the intriguing dilemmas of causality towards *a higher methodological level or by turning to a meta-theoretical 'interpretationist' approach.* We may and should interpret the Kantian approach in general (see my 2004 in 2007, chap. 2) and also his causality theory indeed as a *methodological* one: Kant was basically a *methodological* thinker though he exaggerated the generalization of his model as being universally deduced or derived a priori from (the logical forms of the) Understanding and Reason. To my mind (see my 2006 and 2007) Kant should indeed be understood or/and so to speak 'modernized' and 'liberalized' as an early *methodological interpretationist,* even if he himself did possibly not see his approach that way. This certainly applies also to his theory of causality.

Thus, generally speaking the problems of causality concepts and their pluralities and the integration can be 'solved' or at least successfully addressed by consequently *going methodological and meta-theoretical* in our philosophy of science and epistemology. I think we can even speak of a *methodological turn*, as sketched out earlier (since my 1978, 1993a, b, 2000, 2003 etc.) and above. Methodological scheme-interpretationism can be successfully applied to causality concepts, their plurality problems – including mental causation (see my 1999, 2004, chaps. 6.1, 7, 7.1, 12, 13).

8.3 Woodward's Pragmatical, Action-Oriented and Interventionist "Manipulability" Approach

The standard work *Making Things Happen* by Woodward's (2003) elaborates his "manipulability conception" of causality and causal explanation. For Woodward (ibid. 10 f) "our interest in causal relationships initially grows out of a highly practical interest human beings have in manipulation and control" of situations, actions, conditions and consequences etc., of practical or even only possible circumstances and outcomes of future events, developments and other phenomena based on some "*invariant relationships*". Woodward's idea is that "one ought to be able to associate with any successful explanation" hypothetical or counterfactual experiments that show us

> "that and how manipulation of the factors mentioned in the explanation (the *explanans....*) would be a way of manipulating or altering the phenomena explained (the *explanandum*)". In addition, the "explanation should be such that it can be used to answer .... a *what-if-things-had been-different question" (w-question,* for short*)*: the explanation must enable us to see what sort of difference it would have made for the explanandum if the factors cited in the explanans had been different in various possible ways."[10]

"Causal claims" tell us "what happens under some (not all) interventions"[11] "in appropriate circumstances" (ibid. 65). That seems to be the gist of the interventional approaches to causality including the "manipulability" approach that Woodward spells out in considerable detail by using classical examples. Later (ibid. chaps. 3, 5) he explicitly turns to "interventions, agency and counterfactuals" as well as a "counterfactual theory of causal explanation". "What the relationships we label

[10] Woodward thinks that basically so-called "pure science" and "applied science" are "deeply intertwined"; both are interested in "representing nature in a way that permits manipulation and control" (2003, 12). He sets up as "necessary and sufficient conditions" for *X* to be "a (type-level) direct cause of *Y* with respect to a variable set **V**" if there is "a possible intervention on X that will change Y or the probability distribution of Y when one holds fixed at some value all other variables Z ..... A necessary and sufficient condition for X to be a .... *contributing cause* .... Is that (i) there be a directed path from X to Y such that each link in this path is a direct causal relationship" and "that (ii) there be some intervention on X that will change Y" with all other variables on the path being "fixed at some value" (ibid. 59).

[11] Obviously, the manipulability theory is another interventionist causal theory according to Falkenburg's list (no. IV or 3.). As all different causality approaches also this one has its difficulties as, e.g., Heidelberger (1992, 144 f.) had already made clear: the rather 'anthropomorphic' interventionism would show "convincingly that a philosophical analysis of our agency/acting and our active interaction with the world cannot dispense with *the* (better, ***a***, HL) concept of causality. But at last it cannot make plausible that manipulability by the human(s) is really constitutive for *the* causal relation" (ital. HL). That is right. However Heidelberger just seems still to hypostatize "*the one*" rather comprehensive "concept of causality" to be found and analyzed to "render us a criterion for the asymmetry of causality that does not anthropomorphize natural science". He did at that time not see the methodological way- out by interpretationism/ perspectivism to distinguish between different "variants of causalities" in the practices of the sciences and other related disciplines and to integrate them again in a more or less loose manner for and in the highly necessary inter- and cross-disciplinary discourses!

causal have in common (and the respect in which they contrast with merely correlational relationships) is that they support (also) *potential* manipulations …" (ibid. 120).

Though some such possible, indirect, hypothetical or clustered "causes" and/or omissions etc. play a role in the assigning of "causality" in the wider sense (ibid. 237), there is "a commitment to some version of realism about causation": He emphasizes that this kind of realism "is *metaphysically modest* and noncommittal": "It requires only that there be facts of the matter, independent of facts about human abilities and psychology" (ibid. 121; my ital.).

By emphasizing interventions Woodward especially stresses pragmatically that the specific relationships between the respective *explanans* and the *explanandum*

> "must be *change-relating*[12]: they *must relate changes in the variables in the explanans to changes in the explanandum variable*, and in particular, they must correctly describe *how* the *explanandum* variable would change under interventions on the *explanans* variables" (ibid. 202, my ital.): "Intuitively, to causally explain a phenomenon is to provide information about the factors on which it depends and to exhibit how it depends on those factors" (ibid. 204).

By the way of finer structural analysis, causal claims are only "explanatory" in virtue of providing the right sort of "counterfactual information" (ibid. 205) – according to counterfactual would-be circumstance(s) generally or reliably answering a "w-question" (i.e., "*what-if-things-had been-different question*").

All that need not even necessarily rely on (natural) laws (like in the DN model) but may instead possibly depend on "stable", if restricted, generalizations or counterfactually supported interventions, actions, historical experiences etc.

Thus the broadly construed "manipulability theory" indeed does assign a rather "prominent" "role" to 'instrumental' success – more so than "many competing theories of explanation". That is seen by Woodward as "a virtue of the manipulationist account: … such an account fits explanatory practice in many areas of science much better than the ontologically oriented alternatives" (ibid. 224). In a sense, we can classify Woodward's theory of causality as an action-oriented "new instrumentalist" *and* "new mechanist" approach.

This approach would definitely cover some action-related or -shaped argumentative systematization types, and is thus important for applied and cross-border sorts of causal "explanations" (and mechanisms; see below). One should however to my mind also consider the *choice of the best or a better fitting model* (after Giere 1988, see below) and adapt it to the extant circumstances of the scientific tradition and

[12] Glennan (2012, 316) sees Woodward's manipulability approach thus as a kind of a mechanistic proposal favoring "direct invariant change-relating generalizations". The new mechanist would only add the very "hierarchical" and systemic "character" of the "intervening mechanisms", mitigate the invariance requirement towards just a sort of stable/relatively robust, "regular" or "repeatable "mechanisms as *systems*" – *that is as* "*organized* collections of parts". Also required would be "to provide a more adequate and general mechanistic account of singular causal sequences … to describe causal processes in terms of the interactions of parts as is done in the systems approach, but to recognize that in many singular chains, the parts are not organized in a stable system" (ibid. 323).

methodology. – However, instead of just "softening" and/or weakening, e. g. the law-likeness criteria of "explanatoriness", it seems to me to be much more fruitful and compatible with the methodological scheme-interpretationist as well as the interdisciplinary and cross-level systematizations – "broadly" speaking "explanatory" – argumentations, if one would not stick just to the one and only relatively "weak" or "softened" model like Woodward's major work as of 2003.

A decade later, in his related follow-up studies Woodward (2013) himself does indeed explicitly proceed to an outright *"new mechanistic" solution*, yet still in a rather restricted and yet 'liberal' or 'graded' version. He stays with the terminology of "mechanistic" (or "mechanical"?) explanations, although that label can generally only be used in a broadened sense that indeed gives rise to notorious misunderstandings.[13] Woodward (2013, 40) cites Duprè's (2013, 112 f) skeptical note that indeed "biological systems – organisms, cells, pathways etc. – are in many ways (only, HL) quite misleadingly thought of as mechanisms". But such a wide (maybe all-too broad) usage is meanwhile rather entrenched so that Woodward (and we all) might stay with it. We have to consider however that one should "restrict" the term in some way 'in order to avoid a limitless scope of "mechanistic explanations"'.

For instance, he does not see the specificity of comprehensive systems- and network-bound as well as *topological explanations* as "mechanist" or "mechanical" ones Woodward (2013, V). His "view" is that

> "mechanistic explanations are most likely to be successful when the systems to which they are applied satisfy certain empirical presuppositions, having to do with the possibility of decomposition into stable[14] intervening links, modularity, fine-tunedness, and so on…" (Woodward 2013, 64).

However, for all explanatory strategies and structures there are "limits on its range of application." Woodward even thinks it is sometimes a matter-of-*more-or-less* ("new") mechanist explanation (ibid. 40 f, III). Therefore, "new mechanist explanations" are *gradual or to be graded* according to the vicissitudes of the field and preconditions etc.

In general, I would like to concur with most of Woodward's respective assessments except the above-mentioned terminological remarks. I would however prefer an effect-processual and actionistic/activistic jargon and also an interpretation of the whole matter in a rather pragmatic sense. That would much better delineate and emphasize the really "producing", or effect-productive, and intervening, character

[13] Instead of the narrow traditional wording of classical mechanics, the expressions 'mechanist', 'mechanisms', (even!) 'mechanical', etc. are used in genetics, molecular and neurobiology and -physiology and for organic chemical reactions and systems, not to speak of electrical and electronic or even quantum (mechanical) changes, processes and systems. 'Mechanism' seems to cover such a wide variety of processes that, if undifferentiated, it comes down to be a cover-all umbrella term or "misnomer". It would have been less irritating to talk of whatever 'directed' or dynamical etc. processes.

[14] The well-known systems theoretical concepts and strategies of ultra- and multi-stability are not mentioned, although many, if not most, biological systems and organisms with their notable internal and external flexibility and adaptability, even robustness, are essentially dependent on those capacities.

of bringing about the *explanandum* result and statement. And it could be better combined with the possibility of neatly including action results, technological processes, changes, progress and technological explanations etc. (see my 2007, chap. 7, and 2017, chap. 8).

8.4 The "New Mechanism"

Now, as I understand the so-called "New Mechanism" of the last two decades, that approach does not just restrict itself on applying and projecting schemes or/and models, but it tries to pinpoint "*mechanisms*" (processes in a wider sense, even at times statistical ones) that actually 'bring about', '*produce*' etc. the resulting event or system-integration of processes (so to speak, the traditional *explanandum* in point) in an explanatorily relevant manner by relying on conducive "mechanisms" or inside processes (Craver and Tabery 2015, 2.5). The new "idea of mechanism is a central part of the explanatory ideal of understanding the world by learning its causal structure" (ibid. 2.5.2).

The assets of the new mechanism orientation are, next to an indirect realist interpretation of the old and "new" mechanisms, primarily the studies of the very *conducive materials 'producing' and intrinsic processes of causal (incl. statistical) "mechanisms" of chaining and self-organizing networking*. The analysis of *really effect-generating processes, emerging products and enabling "propensities" are to answer and describe the very intriguing internally differentiated "how come"-questions* by pinpointing to the *carrier processes* in a possibly rather differentiated in-depth analysis.

Active modeling instead of just passive or merely cognitive systematizing seems to be the slogan of the new field: (Only*) productive/result-engendering action paves the ways* of the "new mechanists".

There is no generally accepted definition nor a unique explication of the concept of "new mechanism". Even not as regards necessary and sufficient conditions for delineating and using the concept do we find a common agreement. We just have some "qualitative descriptions" and characteristic features or essential traits that were "most commonly cited" (Craver and Tabery 2015, 2):

- "Mechanisms are entities and activities organized such that they are productive of regular changes from start or set-up to finish or termination conditions" (Machamer et al. 2000: 3).
- "A mechanism for a behavior is a complex system that produces that behavior by the interaction of a number of parts, where the interaction between parts can be characterized by direct, invariant, change-relating generalizations" (Glennan 2002: S344).
- "A mechanism is a structure performing a function in virtue of its component parts, component operations, and their organization. The orchestrated functioning of the mechanism is responsible for one or more phenomena" (Bechtel and Abrahamsen 2005: 423).

These characterizations contain four basic features: A "new mechanism" is a phenomenon (a system of processes, activities, behavior etc.); it has parts (components or sets of some such, specified or fixed boundaries), causal processes, chains and/or networks etc., and organization in the form of 'underlying' lower-level mechanisms, some "stability of process structures or regularity" (ibid.)

The causality-bound mechanisms may display a "conserved quantity account" (as, e.g., a mark transmission à la Salmon, see above). Mechanistic accounts (especially lower level mechanisms) are "truth-makers" for the causal statements or claims. A production-"activity account" (causation understood in terms of productive activities or changes), also counterfactual accounts (as in analytic philosophy, also in manipulative/interventionist approaches) may be conducive, too. Varieties of assemblage/*organization* are spatial-temporal, and mostly according to (sub)fields like electrical/electronical, traditional mechanics or bio-disciplines and/or methods (descriptive, mathematical, dynamical etc.). Even "computational" or "computing mechanisms" are understandably very important in the field. – Usually, however, mere aggregation without methodical ordering is not considered sufficient. Yet emergence/emergent higher level phenomena, interdependence between parts and modularity are indeed typical.

Many mechanists emphasize the hierarchical organization of mechanisms and the multilevel structure of theories in the special sciences (see especially Craver 2007, Ch. 5).

Representatives of the new mechanism would usually focus almost exclusively on etiological and causal relations. However, the new emphasis on mechanisms in biology and the special sciences also demands an analysis of "mechanistic relations *across levels* of organization" (Craver and Tabery 2015, 2.4.5, my ital.).

Not only higher-level organization is common, especially in rather complex systems, but also at times *higher-level* theories (*meta-theories*) and *meta-* and *trans-disciplines* or even *meta-language* approaches for methodological analyses etc. (see my multi-level scheme-interpretationist methodology above and my 2017). So, studies of meta- and cross-disciplinary provenance are a real desideratum for the research field of the "new mechanism" in general and especially in the biological and also any cross-disciplinary sciences.

Craver and Tabery (ibid. 2.5) list a number of characteristics that are *not* necessary nor even typical for new mechanisms. Such mechanisms need ***not*** be deterministic, reductionist, nor be machines or strictly machine-like, linear or 'sequential', localizable, "limited to push-pull dynamics" nor to "fictions or metaphors": to note, "components of mechanisms" can be "widely distributed", and mechanisms might have a "productive aspect", i.e., practically engender or "doing something" (whatever *that* means! Are they really '*agents*'?)

Neither "fundamental laws" nor fundamental causal relations are *per se* considered to be mechanisms. The same is true for logical and mathematical, correlations, entities, objects, symmetries, as well as even inferences, reasons, and arguments (ibid. 2.5.2).

"New Mechanists" rightly criticize the covering-law model of explanation for *not* answering "*how come*"- and deep(er) procedural *why*-questions of events and occurrences.[15] They emphasize with Craver-Tabery (ibid. 3):

> "Mechanists, in contrast, insist explanation is a matter of elucidating the causal structures that produce, underlie, or maintain the phenomenon of interest. For mechanists, the philosophical problem is largely about characterizing or describing the worldly or ontic structures to which explanatory models (including arguments) must refer if they are to count as genuinely explanatory. A rainbow, for the mechanist, is explained by situating that phenomenon in the causal structure of the world; the explanation is an account of how the phenomenon was produced by entities (like rain drops and eyeballs) with particular properties (like shapes and refractive indices) that causally interact with light propagating from the sun. Mechanists typically distinguish several ways of situating a phenomenon within the causal structure of the world."

Most mechanists recognize primarily two main aspects of mechanistic explanation: etiological and constitutive ones.[16] Etiological explanations would according to Salmon (1984) reveal the causal history of the *explanandum* phenomenon, as when one says a virus explains a disease. Constitutive explanations, in contrast, can explain a phenomenon by describing the mechanism that underlies it, as when one says that brain regions, muscles, and joints explain reaching with one's arms. Both forms, but especially the first one, are widely used by the new mechanists nowadays.

A mechanical model (Glennan 2005), i.e. a "description of the mechanism's behavior" and its mechanism(s) that "account(s)" for that behavior, can usually be represented in different ways. For instance, Giere (1988, 78 ff) thinks that you have to "look first for the models" (ibid. 89) and their roles in technology to understand and indirectly connect/link theories with "reality". Technology "provides the connection between or evolved sensory capacities and the world of science; important 'background knowledge' is "*embodied in* the *technology* used in performing experiments" (ibid. 138,140, my ital.). The models are "evaluated in terms of their ability to predict the features" of the respective phenomena and "in terms of the mapping between items in the model and the entities, activities, and organizational features in the mechanism" (Glennan 2005: 17; Kaplan and Craver 2011). Glennan (2010) yet emphasizes that there is no strict delineation between complete and incomplete models; rather, models are continually in the process of articulation and "refinement". "Whether a model is complete enough is determined by pragmatic considerations." Glennan distinguishes between "*productive causes*" and causally "*relevant higher-*

[15] Already decades ago authors like Bromberger, Scriven and Salmon and many others criticized the static subsumption under all-too-general laws for its lack of detailed causation descriptions: "Philosophical arguments against the covering law model often focused on its inability to deal with causal, etiological explanations. The model failed to deliver the right verdict on a variety of problem cases precisely because it attempted to provide an account of explanation without any explicit mention of causation" (Craver-Tabery after Salmon 1984). (New mechanists even extend such criticisms to the covering law model of intertheoretic, micro-reduction.)

[16] Salmon (1984) describes them as two different ways of situating an *explanandum* phenomenon in the causal nexus (see also Craver 2001; Glennan 2009).

level systemic properties" "not caused by the interaction" of the system's parts, "but are constituted by them" (my ital.):

> "Experiments help us to understand the organization of the mechanisms that determine the causal relationships between events, but it is the organization and operation of the mechanisms themselves, not the experiments that make certain properties causally relevant" (2010).

> *"The virtue of the mechanical approach is again that it tells one something quite general about causes – that causes and effects will generally be connected by intervening mechanisms"* (Glennan 2012, 322, my ital.).

And all that would be brought about *either by lawful, systemic, productive or by somehow stable or even counterfactually explainable and/or by "tokened" mark- or event-continuing mechanisms.* (See also Glennan 2017, and note 14 above.)

Darden's distinction between *mechanism schemas* and *mechanism sketches* (Darden and Cain 1989; Darden 2002) is especially interesting in the light of scheme-interpretationism) spelling out corroborated or successful applications of models and just hypotheses with gaps and guesses.

"In discovering a mechanism, it is often crucial to identify gaps that have to be filled in one's model. While no model is ever complete in the absolute sense, some models have lacunae that must be filled before the model is complete enough."

"Mechanism *schemas"* are "abstract descriptions of mechanisms that can be filled in with details to yield a specific type or token mechanism" (ibid.).

Also Machamer et al. (2000, 15) define

> "A *mechanism schema* is a truncated abstract description of a mechanism that can be filled with descriptions of known component parts and activities".

Thus, e.g., the schema: DNA → RNA → Protein can be filled in with a specific sequence of bases in DNA, its complement in RNA, and a corresponding amino acid sequence in the protein. The arrows can be filled in, showing how transcription and translation work. A mechanism *sketch* is an incomplete representation of a mechanism that specifies some of the relevant entities, activities, and organizational features but leaves gaps[17] that cannot yet be filled.[18]

---

[17] "Black boxes, question marks, and filler-terms (such as 'activate', 'cause', or 'inhibitor') hold the place for some entity, activity or process yet to be discovered. The distinction between sketches and schemas is a matter of completeness: schemas are more complete than sketches in the sense that a sketch omits one or more stages of the mechanism that have to be understood if one really wants to solve one's discovery problem" (Craver and Tabery 2015, 3.3).

[18] "Mechanists also emphasize the distinction between a *how-possibly* schema and a *how-actually-enough* schema (Craver and Darden 2013). A 'how-possibly schema' describes how entities and activities *might* be organized to produce a phenomenon. "It is a hypothesis about "how the mechanism works. Such models might be true (enough) or false. A true (enough) 'how-possibly model' is (though we may not know it) also a' how-actually (enough) model'. A 'how-actually-enough schema' describes how entities and activities are in fact organized *to produce* the phenomenon. The term 'how-actually-enough' captures the idea that the requisite 'accuracy' of a mechanistic model can vary considerably from one pragmatic context to another" (Craver and Tabery 2015, 3.3.1). 'How possibly models' are merely 'just-so-stories' similar to promising 'hypotheses' that may fail. They may be fruitful as a possibility but need not necessarily be endorsed.

In general, it is by way of the interaction and description of what usually is called "causal impacts" that sciences get their practice-orientation, as Machamer et al. argue (2000, 1).

However, it is true that "the interactionist's reliance on laws and interactions seems" in some sense to leave out the "productive nature of activities" (ibid.). Instead, like my earlier action-interpretative approach (as of 1978 ff.), these later authors do indeed primarily emphasize the *activities* in mechanisms:

> "The term 'activity' brings with it appropriate connotations from its standard usage", but they use it "as a technical term": "Activities are the producers of change. They are constitutive of the transformations that yield new states of affairs or new products. Reference to activities is motivated by ontic, descriptive, and epistemological concerns":
>
> "Our way of thinking emphasizes the *activities* in mechanisms". "Mechanisms are sought to explain how a phenomenon comes about or how some significant process works. Specifically: Mechanisms are entities and activities organized such that they are productive of regular changes from start or set-up to finish or termination conditions". – "Descriptions of mechanisms show how the termination conditions are produced by the set-up conditions and intermediate stages. To give a description of a mechanism for a phenomenon is to explain that phenomenon, i.e., to explain how it was produced. Mechanisms are composed of both *entities* (with their properties) and *activities*. Activities are the producers of change. Entities are the things that engage in activities. Activities usually require that entities have specific types of properties" (Machamer et al. 2000, 2 f).

An example from neurology might illustrate that (after Machamer et al. 2000, 3):

> "The neurotransmitter and receptor, two entities, bind an activity by virtue of their structural properties and charge distributions of a DNA base and a complementary base hydrogen bond because of their geometric structures and weak charges. The organization of these entities and activities determines the ways in which they produce the phenomenon. Entities often must be appropriately located, structured, and oriented, and the activities in which they engage must have a temporal order, rate, and duration."

Two neurons indeed have to be spatially adjacent to each other to allow for the diffusion of the neurotransmitter.

Thus, according to Machamer et al. (ibid.) "mechanisms are regular in that they work always or for the most part in the same way under the same conditions. The regularity is exhibited in the typical way that the mechanism runs from beginning to end; what makes it regular is the ***productive continuity*** between stages."

However, neither the concept of a clear natural law nor the postulating of new entities etc. would solve the problems of an in-depth description of a rather complex bio- or neurophysiological system, since there are mostly no clear-cut laws of textbook character nor as a rule individually identifiable entities.

But a hierarchical diagram may somehow depicture the system and its inner connections – including usually a bottoming-down arrangement ending at some "lowest" elements ("bottoming-out"). Thus, a diagram might provide some first overlook as a preliminary stage of a sort of pseudo-"explanation" with a sort of hierarchical "adequacy". (Schemes of) mechanisms, i.e. Mechanism schemata, as well as descriptions of particular mechanisms, often take and play the role that is traditionally fulfilled by theory (Machmer et al. 16) and change of theories.[19]

"Entities and activities at multiple levels are required to make the explanation intelligible. The entities and activities in the mechanism must be understood in their important, vital, or otherwise significant context, and this requires an understanding of the working of the mechanism at multiple levels." – "Higher-level entities and activities are thus essential to the intelligibility of those at lower levels, just as much as those at lower levels are essential for understanding those at higher levels. It is the integration of different levels into productive relations that renders the phenomenon intelligible and thereby explains it" (Machamer et al. 2000, 23).

Here indeed, now the meticulous *distinguishing and differentiating of levels and meta-levels* come in the play. Not only in the biological sciences, but rather generally in inter- and cross-disciplinary, even multi-disciplinary research fields and projects the methodological problems of higher-level mechanisms, identification of emergent phenomena and interpretative constructs as well as even "cross-level mechanism sketches" may become really important. Thus far, we have no extant meta-theory to tackle these problems of not only interdisciplinary but also cross-level and meta-level mechanisms.

8.5 From Mechanism Sketches Toward Mechanism Schemata, Systems and Hierarchical Diagrams

In their ground-breaking article Machamer et al. (2000) developed the idea that it is the very mechanisms that in many (notably biophysical) sciences mechanisms as "material carriers" are really doing the groundwork of material concatenations or mark-transmissions (à la Salmon) in on-going causal processes. These "new" mechanisms would indeed more successfully answer some intriguing "how-come"-questions occuring in causal chains, raised for branching processes and networks. Indeed, processual continuity (even if on a higher level of abstraction) and neatly adjacent, so to speak "chaining", activities may better describe and in a sense "explain" the causal history of the "explanandum phenomenon" than, e. g., a traditional covering- law deductive explanation (Craver and Tabery 2015, 3.1).

As regards the phenomena of branching(s) and levels as well as meta-levels of concrete vs. abstract mechanisms we may direct our attention to the Darwinian theory and model of species evolution. Many new mechanists claim that undeniably

[19] "Theory change in biology would be most accurately characterized in terms of the gradual and piecemeal construction, as in turn the evaluation and revision of multi-level mechanism schemata" "Elimination or replacement should be understood in terms of the reconceptualization or abandonment of the phenomenon to be explained, of a proposed mechanism schema, or of its purported components. This contrasts with the static two-place relations between different theories (or levels) and with the case of logical deduction…. This model cannot accommodate the prevalent multi-level character of explanations in our sciences" (Machamer et al. 2000).

biological evolution qua natural selection would be "a mechanism", even if the process of "production" by chance mutations frequently fails (or fails more often than "succeeds") and would not reproduce the same result if started anew after having been rewound (St. Gould).

Therefore, some new experimentalists later on included "stochastic mechanisms" or such "accounts" of them. Only then they could "accommodate the idea that natural selection is a mechanism" (see Craver and Tabery 2015, 2.1.2, 2.6). Doubtless, chance plays a role in evolutionary selection processes; however biologists want to be able to talk about the evolutionary process(es) *as a mechanism*. (Therefore they tend to include and acknowledge stochastic mechanisms – and in quantum mechanics and the notorious measurement problems we have stochastic mechanisms galore!).

However, there is a notable and methodologically interesting point of abstraction and or levels of mechanisms. According to Craver-Tabery (ibid. 4.2)

> "decomposability of mechanisms is directly related to the idea that mechanisms span multiple levels of organization. The behavior of the whole is explained in terms of the activities and interactions among the component parts. These activities and interactions are themselves sustained by underlying activities and interactions among component parts, and so on":
>
> "In short, to say that something is at a lower mechanistic level than the mechanism as a whole is to say that it is a working part of the mechanism. Though the term 'level' is used in many legitimate ways, levels of mechanisms seem to play a central role in structuring the relations among many different models in contemporary biology (e.g., between Mendelian and molecular genetics" (Craver and Darden 2013), between learning and memory and channel physiology (ibid.; Craver 2013), and between population-level variation and developmental mechanisms (Craver and Tabery 2015).

One implication of this view of levels, combined with certain familiar assumptions about causal relations, is that there can be no causal relationships between items at different levels of mechanisms.[20] Claims about cross-level causation, which are ubiquitous in the scientific literature, are best understood either as targeting a different sense of levels or, concerning levels of mechanisms, as expressing hybrid claims combining constitutive claims about the relationship between the behavior of the mechanism as a whole and the activities of its parts, and causal claims concerning relationships between things not related as part and whole (Craver and Bechtel 2007).

If so, then, e.g., natural evolution is a higher-order or abstract *mechanism* (of a partly stochastic provenance). Is it now a "*meta-mechanism*" – or is it just to be analyzed on a higher meta-theoretical level? (By the way, Craver–Tabery mention

[20] "There can be causal relationships between things of different sizes, and there can be causal relationships between things described in very different vocabularies; but (again, conjoined with certain assumptions about the temporal asymmetry of cause and effect and the independence of cause and effect) there cannot be causal relationships between the behavior of a mechanism and the activities of the (*very*, HL) *parts that jointly constitute that behavior*" (Craver and Tabery 2015, 4.2, my ital.).

"epistemic emergence" as the "inability to *predict* properties and behaviors of wholes" from respective parts (ibid. 4.2, my ital.).) "Is *evolution* just such an epistemic emergence"? Or, as I think, rather the use of a higher-level abstract and generalized comprehensive interpretative construct on a meta-theoretic level? *Many rather comprehensive descriptions and concepts of mechanisms are such higher-level constructs.* Are they thus 'epistemic' or theory-laden mechanisms? Indeed, most descriptions and also methodological concept(tion)s are theory-impregnated, especially if abstract(ed) and constructed (see above, methodological introduction). Therefore, mechanists have carefully to get the interpretation levels of their concepts of mechanisms right, and should meticulously distinguish their different bottom-down or abstraction levels of the respective mechanism concepts they use. There might be influential relationships of part-abstract/whole character that are more of a constitutive character than causal ones – across levels. Between the molecular level and some species-genetic properties there might even occur "causal" (or quasi-causal) decisive cross-level impacts of a lower-level phenomenon (like mutations) that have to be interpreted as a not-only constitutive (but in part causal) relation between lower level mechanisms and a higher level one.

Indeed and in short, the relationships of lower-level and higher-level mechanisms should be(come) a prominent subject of not only practical but also methodological studies.

The latter studies are also an important desideratum for the rather general comparisons between different methodological-epistemological approaches as, e. g., interventionist and actionist ones versus traditional objectivist ones like a/the determinist mechanism. (Both Falkenburg and Woodward explicitly delineate a broader concept of causality and rightly criticize the idea of absolute determinism – and they do have a sort of vision for embedding causal analyses in systems on meta-levels and in higher-order epistemological studies.

In conclusion, the methodological study of the development of the approach of really effect-engendering/productive and continuous (chain or branching) processes, i.e. "new" mechanisms, seems to be promising for dealing with describing and answering the very differentiated "how come"- and why-questions of the "new" mechanisms and traditional and other forms of causing.

References and Literature

Bechtel, W., and A. Abrahamsen. 2005. Explanation: A mechanistic alternative. *Studies in History and Philosophy of the Biological and Biomedical Sciences* 36: 421–441.

Beebee, H., C. Hitchcock, and P. Menzies, ed. 2012. *The Oxford Handbook of Causation*. Oxford: Oxford University Press.

Cartwright, N.D. 1983. *How the Laws of Physics Lie*. Oxford: Oxford University Press.

———. 1989. *Nature's Capacities and Their Measurement*. Oxford: Oxford University Press.

Craver, C.F. 2001. In *Structures of scientific theories in Blackwell Guide to the Philosophy of Science*, ed. P.K. Machamer and M. Silberstein, 55–79. Oxford: Blackwell.

———. 2007. *Explaining the Brain*. Oxford: Clarendon Press.

———. 2013. Functions and Mechanisms: A Perspectivalist View. In *Functions: Selection and mechanisms*, ed. P. Huneman, 133–158. Dordrecht: Springer.
Craver, C.F., and W.M. Bechtel. 2007. Top-down Causation Without Top-down Causes. *Biology and Philosophy* 22: 547–563.
Craver, C.F., and L. Darden. 2013. *In Search of Mechanisms*. Chicago: Chicago University Press.
Craver, C., and J. Tabery. 2015. Mechanisms in Science. In *Stanford Encyclopedia of Philosophy*, ed. E.N. Zalta. https://plato.stanford.edu/entries/science-mechanisms [07.04.2017].
Darden, L. 2002. Strategies for discovering mechanisms: Schema instantiation, modular subassembly, forward/backward chaining. *Philosophy of Science* 69 (S3): S342–S353.
Darden, L., and J. Cain. 1989. Selection type theories. *Philosophy of Science* 56: 106–129.
Dupré, J. 2013. Living causes. *Proceedings of the Aristotelian Society Suppl.* 87: 19–38.
Falkenburg, B. 2012. *Mythos Determinismus*. Berlin/Heidelberg: Springer.
Giere, R.N. 1988. *Explaining Science*. Chicago: Chicago University Press.
———. 2006. *Scientific Perspectivism*. Chicago: Chicago University Press.
Glennan, S.S. 2002. Rethinking Mechanistic Explanation. *Philosophy of Science* 69(S3): S342–S353.
———. 2005. Modeling Mechanisms. *Studies in History and Philosophy of Biological and the Biomedical Sciences* 36: 375–388.
———. 2009. Productivity, relevance and natural selection. *Biology and Philosophy* 24: 325–339.
———. 2010. Mechanisms, Causes, and the Layered Model of the World. *Philosophy and Phenomenological Research* 81: 362–381.
———. 2012. Mechanisms. In *The Oxford Handbook of Causation*, ed. H. Beebee, C. Hitchcock and P. Menzies, 315–325. Oxford: Oxford University Press.
———. 2017. *The New Mechanical Philosophy*. Oxford: Oxford University Press.
Heidelberger, M. 1992. Kausalität. *Neue Hefte für Philosophie* 32 (33): 130–153.
Kant, I. 1956[1781/1787]. *Kritik der reinen Vernunft*. Ed. R. Schmidt. Hamburg: Meiner.
Kaplan, D.M., and C.F. Craver. 2011. The explanatory force of dynamical models. *Philosophy of Science* 78: 601–627.
Kim, J. and E. Sosa. 1993. *Supervenience and Mind: Selected Philosophical Essays*. Cambridge: Cambridge University Press.
Lenk, H. 1968. *Kritik der logischen Konstanten*. Berlin/New York: De Gruyter.
———. 1978. Handlung als Interpretationskonstrukt. In *Handlungstheorien interdisziplinär 2(1)*, ed. H. Lenk, 279–350. Munich: Fink.
———. 1993a. *Interpretationskonstrukte*. Frankfurt/M.: Suhrkamp.
———. 1993b. *Philosophie und Interpretation*. Frankfurt/M.: Suhrkamp.
———. 1995a. Das metainterpretierende Wesen. *Allg. Zeits. für Philosophie* 20: 139–147.
———. 1995b. *Schemaspiele*. Frankfurt/M.: Suhrkamp.
———. 1999. Humans as Metasymbolic and Superinterpreting Beings. *Evolution & Cognition* 5: 198ff.
———. 2000. Outline of Systematic Schema Interpretation. In: *Contemporary Philosophy.* Proceedings of the Twentieth World Congress of Philosophy, *Volume 8*, ed. D. Dahlstrom, 121ff. Bowling Green: Philosophy Documentation Center.
———. 2003. *Grasping Reality: An Interpretation-realistic Epistemology*. Singapore: World Scientific.
———. 2004. *Bewusstsein als Schemainterpretation*. Paderborn: mentis.
———. 2006. Kant as a Methodological Interpretationist? In *Kant Today – Kant aujourd'hui – Kant heute. Results of the IIP Conference/Actes des Entretiens de l'Institut International de Philosophie Karlsruhe/Heidelberg 2004*, ed. H. Lenk and R. Wiehl, 26-70. Berlin: LIT.
———. 2007. *Global TechnoScience and Responsibility*. Berlin: LIT.
———. 2017. *Scheme Dynamics: Towards an Action- and Operation-oriented Philosophy of Science and Technology*. Bochum/Freiburg: Projektverlag.
Machamer, P., L. Darden, and C.F. Craver. 2000. Thinking about Mechanisms. *Philosophy of Science* 67: 1–25.
Mackie, J.L. 1965. Causes and Conditions. *American Philosophical Quarterly* 2 (4): 245–255.

———. 1974. *Cement of the Universe: A Study of Causation*. Oxford: Clarendon Press.
Marc-Wogau, K. 1962. On historical explanation. *Theoria* 28: 213–233.
Mill, J.S. 1843. *A System of Logic*. London: Parker.
Rumelhart, D.E. 1978. Schemata: The Building Blocks of Cognition. In *Theoretical Issues in Reading Comprehension. Perspectives from Cognitive Psychology, Linguistics, Artificial Intelligence and Education*, ed. R.J. Spiro, B.C. Bruce and W.F. Brewer, 33-58. London: Routledge.
Salmon, W.C. 1984. *Scientific Explanation and the Causal Structure of the World.* Princeton: Princeton University Press.
Sosa, E. 1993. Varieties of Causation. In *Causation*, ed. E. Sosa and M. Tooley, 234–242. Oxford: Oxford University Press.
von Wachter, D. 2009. *Die kausale Struktur der Welt.* Alber: Freiburg and München.
Woodward, J. 2003. *Making Things Happen: A Theory of Causal Explanation.* Oxford: Oxford University Press.
———. 2013. Mechanistic Explanation: Its Scope and Limits. *Proceedings of the Aristotelian Society Supp* 87: 39–65.

Part III
From Physics to Complexity and Computation

Chapter 9
Causal Mechanisms, Complexity, and the Environment

Jan Faye

Abstract Scientists use a plurality of conceptual frameworks in order to explain the phenomena that really matters to them. This situation arises because some sciences study very complex systems. A non-reductive approach to complexity usually relies on notions like supervenience and emergence to characterize what its proponents see as novel and salient features of composite systems. It is those inherent properties that cannot be reduced to properties that belong to the constituents of the system. However, taking notions like supervenience and emergence seriously we introduce ontological levels of reality, which bring in certain problems of their own. In this paper I suggest a different non-reductive approach to complexity in which complex properties of a system are not inherent but occasioned by the system's environment.

9.1 Introduction

A look into the representational practice of science reveals the presence of a wide and diverse range of different ontologies. This is due to the fact that in their theories scientists use distinctive conceptual descriptions designed to encompass their specific domains of research. Without strong empirical evidence none of these scientific representations can be said to have ontological priority. Those who believe otherwise have the burden of proof since the existence of distinct conceptual frameworks is a matter of fact. Nevertheless, we also know that big things are constituted by smaller things. So it might seem reasonable to think that one could explain the behaviour of bigger and more complex things in terms of the behaviour of their less complex constituents. Such a view is still common among many scientists and philosophers.

J. Faye (✉)
Department of Media, Cognition, and Communication, Section for Philosophy, University of Copenhagen, Copenhagen, Denmark
e-mail: faye@hum.ku.dk

B. Falkenburg, G. Schiemann (eds.), *Mechanistic Explanations in Physics and Beyond*, European Studies in Philosophy of Science 11,
https://doi.org/10.1007/978-3-030-10707-9_9

Today, however, only few philosophers or scientists would hold that, say, our knowledge of biological evolution or population genetics is in any way reducible to particle physics. Whatever scientists take to be a proper explanation of their objects of study is determined by dissimilar research interests, which again depend on the various ways different sciences have structured and categorized the world around us. And how they structure and categorize the world relies very much on the way we experience it, which is a combined effect of our sensory perception, bodily interaction, and purposeful experimentation. The outcome is that we often attribute completely novel properties to complex higher-level systems, properties that seem to transcend those of their constituents. At the same time these novel properties are assumed to have causal powers of their own and may even be imagined to influence those lower-level processes from which they emerge.

From a philosophical point of view the multi-categorical approach and research practice we see within the scientific communities can easily be justified if one acknowledges the naturalistic and pragmatic turn in philosophy of science. Scientists use those conceptual distinctions they find most useful for explaining and understanding their experiences within a certain research area. They don't care much about metaphysical matters. But given the plurality of conceptual schemes of theories there is no neat way to specify how these various conceptions are interconnected. Each of them is always grasped with respect to a particular conceptual and explanatory scheme of investigation. If conceptual pluralism within the sciences is taken to indicate that it makes little sense to look for a basic ontological foundation to which all other conceptual descriptions can be reduced, we do not get a philosophical grasp on how bigger things behave just by looking at how they are constituted. The question is therefore whether philosophy can explain why so many sciences operate with a non-reductive approach to their objects that corresponds to the plurality of conceptual frameworks.

The plurality of ontologies seems to be at odds with the fact that nearly every composite system studied by science can be divided into smaller subsystems. Yet, complex systems are attributed properties that we don't find among the properties of their subsystems. In these cases many scientists, even some physicists, believe that we face novel properties that supervene on the properties of the subsystems and allow so-called emergent behavior. Examples of such properties include emergent behavior of the superfluidity of ^{3}He, isomers in chemistry, flocks of birds, or mental states in neuroscience. Emergent properties cannot be reduced to properties of their parts; they are unexplainable in terms of properties of their constituents because they belong to the system as a whole. Indeed, the appearances of so-called "emergent" behavior at higher levels seem to give rise to different scientific practices covering a large range of distinct ontological levels. So by talking about the organization of complex systems as something that supervenes on their constitutive subsystems and by talking about this organization as giving rise to emergent behaviour, we are lured into seeing the world in a vertical perspective in which new ontological levels are defined with respect to the number of irreducible phenomena. It is exactly the same style of thought, but now moving in the opposite direction, we entertain whenever we attempt to reduce properties of higher levels to lower levels.

Now, people often have different stances in mind when discussing "pluralism" in science. We may distinguish between epistemological, conceptual, and metaphysical pluralism. Conceptual pluralism is stronger than epistemological pluralism, but weaker than metaphysical pluralism. I take epistemological pluralism to be the position that claims that the nature of our different cognitive capacities determine the practical use of multiple concepts. Conceptual pluralism is then the position that there is no uniform theoretical explanation of our experiences but that the same phenomenon can be interpreted in the light of different theoretical frameworks. And finally: metaphysical pluralism is the view that the plurality of theoretical interpretations are indications that the nature of reality is ultimately not one but many. Of course, one is immediately led to ask: what does the acceptance of conceptual pluralism imply with respect to the nature of emergent behavior? Should we be realist or antirealist? Is the emergence found at many levels of scientific practice only as a result of the fact that we experience the behavior of complex systems and their constituents differently because we use different conceptual descriptions when we explain their behavior? Or is emergence a result of the fact that the world really consists of different ontological levels at which new properties and non-reductive laws emerge? If we take emergent behavior to be ontologically genuine, we may ask what its causal status is. How do the various levels interact with each other?

So the metaphysical idea that bigger systems are constituted by smaller systems inexorably pushes us in the direction of considering the world in a vertical perspective as consisting of different ontological levels of reality. In this paper, I show that if we take these hierarchies to be objective levels of reality we create serious problems for the understanding of complex phenomena and their behaviour. Instead of taking an ontological stance in favour of real emergent levels and the possibility of downward causation from these higher levels to lower levels, I suggest that the notion of emergence appears to have a much more epistemological and pragmatic foundation. This is due to the fact that everything in the world is interconnected with everything else and that no cognitive representation can cover all aspects of life. So levels of reality and emergence are rather figures of thoughts that automatically come to mind whenever we want to differentiate between various representations. The present challenge is to understand why humans successfully use different conceptual frameworks to represent the world and also may assume that everything has a material origin.

I think it might be more useful for explanatory purposes, and therefore more metaphysically helpful, to think of complex systems and the rise of a plurality of ontologies from a horizontal perspective. In every domain of nature we can make a division between relatively self-contained or self-sustained systems and their environment. So an *environment* is everything spatially outside what we refer to as the system and which can have an influence on that system. I believe this holds for small systems as well as big systems.[1] Indeed, one can call these domains of systems "complexity levels", as long as one does not hold strong commitments to a *real* ontological hierarchy: I suggest that a complex system is partly a self-contained system encapsulated by another self-contained system where the encapsulating system functions as the *environment* of the encapsulated system.

Thus, it is in the causal interaction between a complex system and its environment that some properties of the system may be described as features of emergence. Hence, we may alternatively understand complex system from a horizontal perspective in which so-called *"emergent"* phenomena are regarded as the results of a system's causal interaction with its environment. Thus "emergent" features of a system are not in this perspective brought about by the internal organization of the subsystems alone but rather by the response and feedback from the system's surroundings.

Any system can be described as being encapsulated by an environment of greater complexity and any composite system is more complex than its subsystems because it also functions as the environment of its subsystems. In general, one may say that a system is self-contained or self-sustained if *causal mechanisms* internal to the system determine its surface properties, its limited range of activity, and its organization. However, no system exists in isolation from its environment, and I propose that it is the cognitive division we make between a system and its environment that is seen to give rise to the "emergent" behaviour. My claim is that there is no hocus pocus in explaining so-called "emergent" behaviour with respect to the epistemic and ontological commitments of conceptual pluralism.

9.2 The Vertical Perspective of Explanation

It is commonly said that properties of higher-level systems arise out of the properties and relations that characterize their constitutive elements. Most of these properties of higher-level systems are "resultant" but some are "emergent". The "resultant" properties can be reduced, or so it is assumed, to properties of the subsystems, whereas the "emergent" ones cannot. Some philosophers have even suggested that the whole, which exhibits the emergent behavior, in fact gives rise to a kind of behavior, which results in an efficient downward causation from higher levels to lower levels. I find the concept of efficient downward causation quite dubious, because it is difficult to see how downward causation could meet some reasonable criteria of causation.

Consider a flock of starlings or a school of herrings. They form a complex system moving back and forth in abrupt movements. Each individual starling or herring

[1] It may seem arbitrary to make some form of spatial containment the environmental boundaries of most systems. The gravitational force of a body is not confined to the spatial containment of its mass and one could easily argue that this force, which stretches far beyond the matter itself, belongs to a system together with the body. Nevertheless, although a spatial distinction between body and its environment may be neither sufficient nor necessary for specifying the boundary, our concept of a physical entity is very much associated with what we can and cannot count, such that spatial concentration of matter for many purposes defines a body's boundary to its environment.

behaves separately as a member of a flock or a school or as an individual bird or fish. Flying together in a flock the starlings create a complex system, which seems to cause each of them to manifest behaviour that they do not show as single individuals. So the question we should ask ourselves is how we can explain the flock behaviour of each starling. This is how we usually think of it:

(1) *Reductionism*: Flock behaviour can be reduced to the relative behaviour of each individual starling and this behaviour can again be reduced to what is going on in the neuronal system of each starling, and so on.
(2) *Emergentism*: Flock behaviour is an "emergent" property that supervenes on a composite system made up of the relative behaviour of each individual starling forming the flock.

Indeed, reductionism is one possibility which still has its proponents, and in some areas of science this view has been quite successful. However, what is called the new mechanistic approach to complex system is a way to provide the details for emergentism although this was not the original intend behind the approach (Machamer et al. 2000). Both reductionism and emergentism are what I call vertical approaches oriented towards opposite metaphysical goals of research.

The train of thought that leads to the vertical approaches cannot be better exemplified than by Oppenheim and Putnam's argument for the unity of science. In 1958 they set up a picture obeying two principles: (1) *the principle of evolution* and (2) *the principle of ontogenesis* (Oppenheim and Putnam 1958). The first principle states that for every level of organization in the universe there exist at a certain time t stages of the evolution only up to a particular level of organization, n, and no higher level, $n+1$. The second principle maintains that for each level of organization n there is a time of the evolution, say, $t-1$ when n did not exist, but when all its components existed at level $n-1$. From those two principles they concluded that everything at a certain level is ontologically dependent on the level below. This claim may be called *ontological physicalism.* However, Putnam and Oppenheim didn't know what we now know. Nevertheless, their basic idea can be illustrated with the physics we know today. Shortly after the fundamental particles like quarks, leptons, and gluons were created in the very early universe, these particles merged together and formed protons, neutrons, etc. As the universe evolved by expansion, protons and electrons made up hydrogen atoms, which later combined into molecules, etc. Eventually stars formed and more complex atoms and molecules were made out of which biological material came into existence.

The O-P picture can be combined with either the reductionist approach or the emergentist approach. Complex systems form a hierarchy of subsystems due to the evolution of the universe, and each of these systems are constituted by its components that exist on a level below. In spite of the fact that Oppenheim and Putnam defended ontological reductionism and used the principles to argue that everything at a higher level in principle could be reducible to a lower level, it seems evident that the principles in isolation do not answer why every feature of a certain level can be explained by everything that exists below this level. It seems as if they never considered emergence a possibility.

Since reductionism has not been overwhelmingly successful everywhere, many philosophers and scientists, especially those interested in the life sciences, have given up on this approach and have introduced the idea of emergence, following earlier ideas of John Stuart Mill and C.D. Broad. According to the present notion of emergence, a system consists of causal mechanisms whose activities as a whole constitute "emergent" phenomena. Flocks of starlings form a complex system made of thousands and thousands of starlings. The entire flock behaves very different from how a single starling would do. The reason is that the activity of the mechanism of the complex system as a whole, and whereby the "emergent" flock behaviour of the individual starling, owes its existence to the property of the mechanism of the flock. Thus, the dynamics of the flock as a whole seem to influence the behaviour of each single starling. If this is so, we seem to have to introduce a case of downward causation. A property of the entire flock causes a property of each single starling. Can this be true?

Most definitions of mechanisms agree on a minimal set of conditions that must be met for some system to form a mechanism. It must contain *parts, activities, organization,* and a *phenomenon*. Thus a complex system consists of different elements that act together such that their mutual actions establish an organization which then gives rise to emergent properties of the system. Hence, we can characterize a complex system in the following way:

> A complex system S consists of the elements $x_1, x_2, x_3, \ldots, x_n$, performing the activities, φ_1-ing, φ_2-ing, φ_3-ing, ..., φ_n-ing, as well as an organization of these activities in such a way that the causal behavior of them as a whole *constitutes* the phenomenon, ψ-ing, which is the emergent behaviour of the composite system S.[2]

In other words, the mutual flock-behaviour (the "emergent" phenomenon we want to explain) of all the starlings is explained by saying that it is constituted by the mechanism that characterizes the flock of starlings as a whole.

The important relation here is the constitution: the activity of the mechanisms as a whole is said to constitute the phenomenon. Constitution, in contrast to identity, is an asymmetric relation between the parts of the system and the system as a whole. One possible way to specify such a relation would be to say

> *Constitution*: A set of elements $x_1, x_2, x_3, \ldots, x_n$, displaying φ_1-ing, φ_2-ing, φ_3-ing, ..., φ_n-ing, *constitutes* the ψ-ing of a complex system S at time t if, and only if, (1) $x_1, x_2, x_3, \ldots, x_n$ and S exactly coincide at t, and (2) the activity ψ-ing of S would not have existed, unless the activity φ_1-ing, φ_2-ing, φ_3-ing, ..., φ_n-ing of $x_1, x_2, x_3, \ldots, x_n$ had existed, whereas some activity other than φ-ing activities of $x_1, x_2, x_3, \ldots, x_n$ might have existed without any ψ-ing activity of S had existed.

Applying this definition to our example of flock behaviour, we see that φ_1-ing, φ_2-ing, φ_3-ing, ..., φ_n-ing stand for the activities of the individual starlings $x_1, x_2, x_3, \ldots, x_n$. At the time t every single starling becomes parts of organized flock such that the causal behavior of performing φ_1-ing, φ_2-ing, φ_3-ing, ..., φ_n-ing taken together constitutes the flock behaviour ψ-ing. It is obvious that there would be no flock

[2] This is essentially Craver's formalism. See Craver (2007).

behaviour without the activity of individual starlings, but each individual starling may display other sorts of activities without showing flock behaviour.

What then is φ_1-ing, φ_2-ing, φ_3-ing, …, φ_n-ing and ψ-ing more specifically standing for in the example of starlings? Let us assume that ψ-ing refers to the density variation of the flock and that φ_1-ing, φ_2-ing, φ_3-ing, …, φ_n-ing denote the flight of each starlings x_1, x_2, x_3, …, x_n. Science has discovered that in order to describe the flight of a starling in a flock it is possible to establish a model that indicates how φ_1-ing, φ_2-ing, φ_3-ing, …, φ_n-ing may be organized. Basic models of flocking behavior suggest that the starlings are controlled by three simple rules of organization: The first is keeping a distance to the other starlings, also called *separation.* 'Avoid crowding neighbors' is a rule that can be described as short range repulsion. The second is about orientation. 'Steer towards the average heading of neighbors' is another rule called alignment. The third one is called *cohesion.* It concerns that the distance to some of the other starlings should not be too great. 'Steer towards the average position of neighbors' can be described as long range attraction. One of these models, called topological distance model, assume that each starling pays attention to 6 or 7 of its closest neighbors. By the help of the three simple principles of organization, scientists are able to simulate the flow movement in a very realistic way, creating complex motion and interaction that would be extremely hard to create otherwise.

To simplify the discussion let *A* designate the total set of activities, φ_1-ing, φ_2-ing, φ_3-ing, …, φ_n-ing, that are associated with the behaviour of all *n* starlings, and let *B* denote one of the flock properties ψ-ing such as spreading out. The emergent properties, *B*, can then be said to supervene on the behavior of the mechanisms of the flock as a whole, *A*, if and only if *B* is distinct from *A*, but *B* is determined by *A*.

> *Emergence*: The activity *B* is an emergent property of an organizational composite system *S*, if and only if (i) *S* has it, but *B* does not belong to the activities *A* of the constituents x_1, x_2, x_3, …, x_n, and (ii) each change of *B* corresponds to a change in the mutual arrangement of the activity of the parts, *A*, but not vice versa; i.e. the activity of some of the constituents may change, and therefore *A* alters, without the activity of the flock *B* is altered.[3]

In other words, but briefly put, *B* is an emergent property of *S*, if and only if x_1, x_2, x_3, …, x_n and *S* coincide in space and time, and at each and every activity of x_1, x_2, x_3, …, x_n is a property of *S*, whereas each and every property of *S* is not a property of x_1, x_2, x_3, …, x_n. The consequence is that the flock behavior of the entire flock, *S*, according to the new mechanism approach, is not *identical with* the behaviour of each individual starling. The mechanism of their behaviour taken as a whole constitutes the flock behaviour but is not identical to it.

---

[3] Indeed, this is a week conception of emergence that even reductionists concerning explanation might accept, because it does wave aside a possible explanation of how the individual starlings produce the behaviour of the flock. The conception has been chosen deliberately in order to include both a supervenient description of the flock behaviour and yet allow an explanatory reduction of this behaviour in terms of its subvenient basis. So a non-reductionist who insists on the existence of genuine levels of reality has to argue in addition that such an explanation is not possible because it conceptually presupposes the existence of what has to be explained.

As Jaegwon Kim has observed, and as it is stated in condition (ii), most emergentists accept mereological supervenience: "Systems with an identical total microstructural property have all other properties in common.[4] Equivalently, all properties of a physical system supervene on, or are determined by, its total microstructural property" (Kim 1999). The implication of this condition is that if we have two systems *S* and *S** constituted by exactly the same lower-level mechanisms, they exhibit the same resultant and emergent properties, and nothing but these lower-level mechanisms determine the resultant and the emergent properties. In other words, the emergent properties of a system supervene on the intrinsic mechanisms of the system and only on these mechanisms.[5]

The general suppositions behind the new mechanistic philosophy are that, except for the lowest level that constitutes all other levels, genuine new forms of organization of mechanisms take place at higher and higher levels. These higher-level mechanisms are realized by lower-level mechanisms. Moreover, higher-level mechanisms have "emergent" property or exhibit novel behaviour that cannot be found among the lower-level mechanisms. Given these two assumptions, the proponent of the new mechanistic philosophy seems able to explain why there are so many distinct ontologies in science by pointing out that each new emergent phenomenon takes part in a higher-level mechanism that again provides a new phenomenon which again takes part in even higher-level mechanisms and so forth. Each new mechanism constitutes a genuinely new and irreducible behaviour, which cannot be comprehended in terms of the conceptual framework that has shown its fruitfulness in understanding a lower level. The downside is that the notion of levels of mecha-

[4] Here Kim has a note saying. "Obviously extrinsic/relational/historical properties (e.g., being 50 miles to the south of Boston) must be excluded, and the statement is to be understood to apply only to the intrinsic properties of systems. There is also a tacit assumption that the intrinsic properties of a system determine its causal powers."

[5] In her book *Unsimple Truths* (2009) Sandra Mitchell criticizes Kim for conflating *compositional materialism*, i.e., that everything consists of one physical substance, and *descriptive fundamentalism*, i.e., that there exists a complete and privileged description that can capture everything which can be said about this substance. Mitchell thus argues, "Kim's attempt to clarify the philosophical conception of emergence has stripped it of any scientifically interesting features, and hence it fails to apply to the properties that scientists have identified as emergent, properties like division of labor in social insect colonies, which have different material realizations for different species of ants, bees, and termites, and perhaps the same dynamics of how the higher-level properties are generated" (p. 32). She also points to the flocking behaviour of starlings as an example of emergence, and she regards such behaviour to be an expression of the non-linear dynamic relationship that exists between the individual bird and the collection of birds. I agree that descriptive fundamentalism is unsupported because any description is always partial. Descriptive fundamentalism presupposes that an exhaustive description can be given in terms of the intrinsic properties belonging to the most fundamental level of reality. But Mitchell continues to talk about "higher-level properties" as if the expression "higher-level properties" has a fundamental metaphysical meaning. In contrast, I suggest that the so-called emergent properties from a vertical perspective are *intrinsic* properties of "higher-level" systems, but from an alternative horizontal perspective should be understood as *relational* properties between "lower-level" systems and their environment. Thereby I attempt to show how we can avoid the strong metaphysical commitments that are inevitably associated with the use of expressions like "emergence" and "higher-level properties".

nisms seems to invite the endorsement of either epiphenomenalism or a form of downward causation. [6]

The argument for such an assertion is this: Assume a series of mechanisms, *A*, causes another series of mechanisms to have the property *A**, and two higher-level phenomena *B* and *B** are realized by *A* and *A** respectively, then what are the causal status of *B* and *B**? We can think of the cohesion of each starling becomes too tight, *A*, which causes each starling to separate, *A**, by turning away from its nearest neighbour. On a higher level, what we get is a flock behaviour consisting in an increasing density of the flock all together, *B*, and then a behaviour *B** in which a flock turns north and spreads out. Now *B* may have causal power in its own right or it may be causally impotent. If *B* is *causally impotent* it does not cause *B**, and we have a case of epiphenomenalism, whereas if *B* is *causally efficacious*, we may have a case of downward causation in which *B* has a causal influence not only on *B**, but also on *A**. Either way it creates serious problems for the mechanistic philosophy.

9.3 Some Objections

Here are some objections to the hierarchy model of reality. If we think of *B* as causally impotent it seems that all emergent properties at any higher level have an epiphenomenal nature. But a property without any causal power is epistemically inaccessible and functionally inadequate. The consequence would be that all the properties in question would be inaccessible for the starlings and could have played no causal role in the adaptation of flock behaviour. This is definitely not the case. Rather it seems evident that *B*, which stands for the increasing density of the flock, not only causes the starlings' visual recording of the increasing density, but it may also cause a falcon to be unable to prey on the starlings. On the other hand, if we accept causal closure for the ontological level of *n*, no property of level $n+1$ can have any causally influence on the property of level *n*, i.e. the whole does not causally determine the behaviour of its parts. We cannot have downward causation because downward causation allows for overdetermination in the sense that *A** is not merely caused by *A* but also by *B*. However, it seems evident that we already in the outset of an explanation of the causal mechanisms of flock behaviour appeal to the membership of a flock.

The interesting point is that we cannot causally describe flock behaviour and *why* it takes place, unless we presuppose the existence of certain organizational principles that we attribute to the subsystems but only to the subsystems as parts of a

[6] On a response to this, and to Kim's exclusion problem in particular, see Glennan (2010). His main objection, which I consider as sound, is that it is only objects realizing properties rather than properties themselves that can produce anything. Nevertheless, in my opinion his layered mechanistic solution is not the answer to flock behaviour, since properties of complex systems are not emergent properties of some internal mechanism but the manifestation of such a mechanism caused by something external factor.

bigger system. The organizational principles like separation, alignment, and cohesion explain *how* the flock behaves but not *why* the individual starling behaves as it does as a member of a huge flock. For instance, a single starling or two starlings would not behave according to these principles. Instead, the behaviour of each single starling described by the organizational principles governs the flock behaviour. The conclusion seems to be, if we can generalize from the present example to other alleged cases of emergence, that causal processes within a system determine the emergent features of the system. Nevertheless, these overall features can be causally explained by the action of the subsystems only in case we can attribute certain features of organization to the subsystems in relation to their membership of a larger system. Features such as separation, alignment, and cohesion only make sense to ascribe to a single bird if we already know that this bird is a member of a flock (Faye 2014). It is the ascription of emergent features already to the subsystems, i.e., features they have only as a member of a larger system, which help us to identify the causal processes within this larger system and to describe how these give rise to the emergent general behaviour of the system. Thus, emergent properties function as explanatory constraints for lower-level behaviour. But they also function as ontological constraints: it seems to be the membership of the flock that causes the actual behaviour of each starling to be performed according to the rules of separation, alignment and coherence.

Apparently, the upshot is that we have a case of downward causation between two levels of reality. Some people may doubt that downward causation is really a case of causation. In the non-reductionist scheme of the vertical perspective downward causation has meet all the criteria of causality. There exists the right counterfactual dependency, and we can both produce flock behaviour of a single bird by putting it into a flock, as we likewise can produce single-bird behaviour by separating a single bird from the flock. However, in a reductionist scheme so-called downward causation will be relegated to feedback mechanisms that only involve the properties of each single bird.

9.4 A Horizontal Perspective

A vertical perspective on how everything in nature is built up by smaller and smaller things runs into metaphysical predicaments with either reductionism or emergentism. It assumes that so-called emergent properties of bigger systems are either reducible to intrinsic properties of smaller systems or they supervene on intrinsic mechanisms of smaller systems. Neither of these positions seems to be in harmony with the epistemological fact that we use a manifold of explanations in the natural sciences. But there is an alternative, I believe, that provides us with a much more satisfactory account. This is the horizontal perspective in which every system is taken to be surrounded by an environment, which is at the same ontological level as the system itself. Likewise, the surrounding environment, including the original system, can be regarded as another and bigger system, which may be surrounded by a further environment, and so forth. Within such a perspective there are causal interactions between systems and their environments, but neither reduction nor emergence nor downward causation between different levels.

What then is the big difference between the vertical and the horizontal perspectives? From a vertical perspective all properties are seen as *intrinsic* (inherent) to the system *S* either in the form of emergent properties or the reductive properties of its parts. In both cases the system has these properties regardless of its environment. Considered from a horizontal perspective a system has both *intrinsic* as well as *extrinsic* properties.[7] The *extrinsic* properties are those properties which a system *S* has by causally interacting with its environment *E*. These properties can neither be said to supervene on the intrinsic properties of *S* nor be said to be reducible to the intrinsic properties of *S*. In case *S* refers to a starling the intrinsic properties are those that characterize its brain processes, whereas its extrinsic properties are those the starling has in relation to the other starlings because it causally interacts with the other starlings. In this case separation, alignment, and cohesion are *extrinsic* properties. In other words, all the other starlings make up each single starling's environment.

Again the flock as a whole may be regarded as forming another system *S** that has both internal and external properties. In this sense you can say that the complex system *S*, i.e. the single starling, is a subsystem and therefore part of a more complex system *S** that consists of all the starlings. In other words, any complex system *S* can be seen as nested by another complex system *S**, where the nesting system *S** works as the environment *E* for the nested system *S*. Now separation, alignment, and cohesion serve as *internal* properties of *S**, and the environment *E** may consists of a landscape and a peregrine falcon that attempts to isolate one of the starlings. The external properties of *S** will in this situation be the behaviour of the flock with respect to the falcon and the landscape below. This possible shift of focus between *S* and *S** offers us an explanation of the existence of different ontologies. It also shows why descriptive fundamentalism fails and why our general epistemic situation in science is contextual.

So we have, according to the horizontal perspective, that any system, big or small, *S*, consists of some mechanisms that possess certain internal properties, *C*, which may causally interact with some entities of its environment *E*, and may thereby cause certain *external* behaviour *B* of *S* with respect to *E*.[8] What is considered to be the causal interaction of system *S* and the environment *E* may likewise be regarded as part of the internal behaviour of mechanisms of a bigger system *S** that

[7] Extrinsic properties are only one kind of external properties. Other external properties are spatial and temporal relations. Likewise intrinsic properties are not the only internal properties. Dispositions would normally also count as internal properties.

[8] Depending on how one considers mechanism we may either say that a system consists of some mechanism or that it contains a mechanism. If we include, as the new mechanistic philosophy does, entities that exhibit a certain activity into the characterization of the mechanism of the system, then the system obviously consists of its mechanism. But if we exclude those entities from the characterization of the mechanism, then it would correct to say that the system contain or exhibit a mechanism.

enclosed S as a subsystem. The interaction between S and E may not only be that S and E affects the other but also that the information about the outputs of S to E is available to S itself via feedback of various sorts. For instance, a starling, regarded as a complex system, is in virtue of natural selection born with behavioral capacities such as separation, alignment, and cohesion that are confined to the information encoded in the neural structure of its brain. Whenever a starling causally interacts with its environment, consisting of six or seven other starlings surrounded by the same number of starlings, it receives information about the position and the flight direction of its neighbors through its visual sense experience at the same time as it receives internal information from its body about its own correction of position and flight direction. Thereby it also starts to show flock behaviour.

Thus, the *capacity* of separation, alignment, and cohesion are *internal* properties of each starling due to some neuronal mechanisms, whereas the *actual* display of these properties are the *external* properties of the starling that are caused by the starling's interaction with the surrounding 6 or 7 starlings. The flock behaviour is caused by the environment via some internal neuronal mechanisms that can be identified as the capacities of separation, alignment and cohesion. Such capacities are dispositions that have to be activated by external signals before any particular starling exhibits flock behaviour. Internal to a starling there exist some neuronal mechanisms that dispose it to behave in a certain way while it receives information about a certain number of other starlings. In general, we can say that a system S has a disposition D to exhibit a behaviour B if and only if S displays B whenever D is activated by the appropriate external stimuli E. Indeed what is appropriate in biology is established by natural selection. Having a disposition is normally assumed to be grounded in a *categorical* basis corresponding to some *intrinsic* mechanisms in the brain. So before D is activated under the influence of E, parts of a system S must be in a state, say C, which is taken to be the categorical basis of D. This categorical basis of D is the internal mechanisms of S organized in an appropriate structure which gives rise to B.

An important part of the horizontal analysis is how C, D, and B are related. The question can be divided into two: (1) what is the relation between a system's disposition D and its categorical basis C? And (2) what is the relation between a system's disposition D and its activation or manifestation, i.e. the exhibition of B? Addressing the first question, the traditional reply comes in one of three. One suggestion is that disposition is *identical* to its categorical basis; another is that a disposition is *constituted* by its categorical basis but not identical to it; and finally a third suggestion that a disposition is *caused* by its categorical basis. I think of all three responses as inadequate as they stand.

The first suggestion implies reductionism. A categorical basis consists of certain intrinsic properties of the system. Therefore if a disposition is identical to its categorical basis, it reduces to some intrinsic properties of the system such as neuron patterns. But this is at odds with how we identify dispositions in relation to their causal role. The second proposal corresponds to be embracing emergentism. Dispositions supervene on the mechanisms of the categorical basis as a whole. This solution is what we want to avoid. Finally, the third suggestion seems to offer property dualism. The categorical basis caused the flock behaviour, but it is difficult to see how the categorical basis should be able to cause the disposition of which it is a categorical basis. Nevertheless, all three suggestions seem to sit well with the vertical approach.

Assuming that all three suggestions are insufficient as they stand we have to look for an alternative. As an alternative I propose a horizontal approach according to which dispositions are regarded as relational properties that cannot merely be grounded in the intrinsic properties of the categorical basis. The categorical basis is, using David Armstrong's terms, states apt for the producing certain ranges of behaviour but also apt for being produced by certain ranges of causes (Armstrong 1968, Ch. 6).[9] A disposition does not refer to its categorical basis itself but to the capacity of the categorical basis to stand in causal relations to other states of affairs. The capability of the categorical basis to act and react consists of information about how to cause an output given the appropriate input – assuming that this information involves the relationship of the categorical basis to other states – then a disposition is identical to the capacity of dealing with such information and not to the intrinsic properties of the categorical basis that sustains information processing. When it comes to human beings, dispositions need neither be apt for being produced by one kind of input only nor be apt for bringing about one kind of output. In the case of starlings we have three different dispositions; separation, alignment, and cohesion. Each of them is defined in relation to the behaviour of a starling whenever it receives perceptual information about the behaviour of surrounding starlings, and each of them can only be identified according to the manifested behaviors, namely moving away from neighbors, moving in the same direction as them, and moving closer to them. The "resultant" behaviour is their flock behaviour.

The "resultant" behaviour corresponds to the "emergent" behaviour of a system in the vertical analysis. However the "resultant" behaviour cannot be reduced to the internal mechanisms of a biological system *S*, like a starling, because the function of internal mechanisms that are concerned with flock behaviour is always understood in terms of its interaction with the environment. Neither does the "resultant" behaviour supervene on some internal mechanisms, since it is caused by the capacity of these mechanics to produce causal outcomes in case they receive information from the environment. A so-called "emergent" property is not an intrinsic property of the system as it seems to be in case we think of complex systems as consisting of hierarchy of layers of subsystems. The "resultant" behaviour is an *extrinsic* property the starling acquires from the surrounding environment. One could also say that the

[9] Already Lewis (1966) talked about the causal role of experience is identical to the causal role of the physical state and therefore experiences must be physical states. But where Armstrong and Lewis were willing to identify an experience with its physical basis, because they believe that the ascription of experience and the ascription of neutral states to a system have the same reference, I for my part hold that the disposition of a system refer to the causal capacity of physical states that originally made the categorical basis an apt state. I take a disposition is not identical to a categorical basis as such but to its capacity to process information in responses to some external stimuli. Thus a behavioral disposition is identical to a neuronal disposition and not to the neuronal basis itself.

"resultant" behaviour is the external properties of the starling that show up in relation to the other starlings. By this relation the starling, which can be regarded as a complex system itself, becomes part of an even more complex system containing all the starlings.

What if we distinguish between manifested and non-manifested dispositions? Would this bring light to our discussion? We may say that any non-manifested disposition D is not *ontologically identical* to the *categorical* basis C but to content of information that can be associated with C, just as the *non-manifested* disposition D is *ontologically different* from the *manifested* disposition, B. So the horizontal response to the first question would be something like this: C and D are not *ontologically identical* in spite of the fact that S loses D if and only if it loses C. It is obvious that the neuronal basis that maintains certain pieces of information cannot go out of existence without that information being destroyed. Furthermore, C and D are *epistemically different* in so far as we cannot identify or recognize C as the categorical basis of D if we did not understand D independently of C. We can identify D only in terms of its manifestation B that does not have C as its categorical basis. The horizontal answer to the second question is crucial: The disposition D becomes manifest whenever the system S stands in the right causal relation to its environment E. The exhibition of B happens once the right input from E interacts with C. D is normally not defined in terms of what activates it, because D itself is regarded as causally separated from the stimuli from E. Activated the non-manifested disposition D results in a manifested state B. It is part of the meaning of a specific disposition D that it realizes or can realize B in virtue of the presence of a particular E. The activation of D is an effect of the system's interaction with its environment and the activation results in B, which is the manifestation of D.

Applying the above analytic scheme to a complex system like a flock of starlings, it is reasonable to presume that the dispositions of starlings in form of separation, alignment, and cohesion have their categorical basis C in particular structures of the neuronal network gained by each starling in virtue of adaptation and preservation. In the case of a starling moving in a flock, the external stimuli from E are its visual image of the distance to and direction of the six or seven surrounding starlings. The capacity of a starling to receive such visual information is a disposition, D_1, and the capacity of behaving in response to this information is another disposition, D_2. But we shall ignore such subtleties for now. Instead we shall present a horizontal explanation of how any so-called "resultant" behaviour of the flock comes about.

The behaviour of the entire flock is the amalgamation of the flock behaviour of each single starling. The flock behaviour of a single bird can be causally explained only if we consider its behaviour in relation to the behaviour of other individual starlings that serve as the environment to this particular one. Hence for each and every starling in a flock all the other starlings form its environment, whereas it may itself be considered as a system. The flock behaviour can be explained due to the interaction of the starlings only in case we can attribute certain causal features of information to the individual starling and certain *manifest* dispositions in relation to their membership of a larger system. The actual behaviour of a particular starling in a flock occurs, because it is surrounded by all the other starlings and it responds to their presence in a particular way. Furthermore, the manifestation of the dispositions such as separation, alignment, and cohesion make sense to ascribe to a single bird only if we already know that it is immersed in an environment in which it is a member of a flock. It is the ascription of the environment to the starlings by which we can distinguish between different forms of behavior and by which we can explain the performance of each single as a member of a flock. Knowing the environment helps us to identify the causal processes that constrain each single bird and thereby to define the content of information that is connected to the categorical basis.

The horizontal perspective transforms every "emergent" property in the vertical perspective, like flock behaviour, into a relational property between the system and its environment. I call such a property an extrinsic property in case it exists because the system stands in a certain causal relation to parts of its environment. A relation can exist either because something outside the system is causally interacting with the system or because something outside the system and the system are connected by more conventional manners. Being a husband and being a wife are relational properties in the latter stipulative sense, but being a cause or being an effect is a relational property in the former sense. If the output from a system is a result of its interaction with something in its surrounding we may call it an extrinsic property.

My claim is that the manifested disposition *B* (the "emergent" phenomenon) should be seen as an extrinsic property of the system *S* brought into existence by its interaction with the environment *E*. An internal structure of the system and an external structure of the environment stand in a causal interrelationship, and it is the system's outputs in reaction to the inputs from the environment that is mistakenly taken to be an "emergent" property. But where does this horizontal approach bring us with respect to an "emergent" property like the density of the flock? It is quite evident that this is an internal property that belongs to the entire flock and not to the behaviour of any single bird. How does such a property arise? If every single starling is heading north and then turns eastward, the entire flock is flying north before going east. Again if each and every starling steers towards the same average position as its neighbors and then repulses away from them, the entire flock condenses for thereafter to separate to avoid some starlings clashing into one another. It is also this behaviour that provides the density of the flock. The density of a flock is a resultant property of the flock that is identical to the position of each starling in the flock.

I have used a flock of birds as a paradigmatic illustration of how alleged examples of emergence can be handled in a horizontal perspective. By considering "emergent" properties to be extrinsic properties that a system has in relation to its environment, we can find them everywhere in nature without being committed to epiphenomenalism or downward causation. An extrinsic property of minerals is their hardness. This property does not belong to their chemical composition and crystalline structure, but defines the mineral's resistance of being scratched. The chemical composition and crystalline structure provides the mineral with a non-manifested disposition to respond in a specific way whenever something interacts with it. The manifest disposition is the reaction the mineral exhibits in relation to the scratching object. Another example is the allotropic nature of phosphorus. Allotropy is the property of some chemical elements to exist in two or more different forms.

The main allotropic forms of phosphorus are white and red phosphorus depending on how the bond of the phosphorus atoms in the molecular is structured. These two forms exhibit very different properties which both seem to be "emergent" in the sense that they are not explainable in terms of the molecular structure itself. Heating white phosphorus may eventually turn it into red phosphorus by changing the molecular structure. This happens because of a changing environment of an increasing temperature. White phosphorus has to be kept under water, since it bursts into flames at about 30 degrees Celsius, while red phosphorus ignites at about 300 degrees Celsius. Again we see that it is events in the environment of phosphorus that provides the various allotropic forms with an extrinsic manifestation of their different dispositions.

9.5 Summing Up

Nothing stands alone in this world. Small things as well as big things are embedded in or surrounded by an environment which has a huge impact on what these things are and how they behave. Still, because larger systems are made up by smaller systems, it is quite natural to think of them as forming a hierarchy of existence where the smallest systems are at the lowest levels and the biggest systems at the highest levels. This figure of thought fits well with what we know about the evolution of the universe. The smaller things like the elementary particles appeared before the atoms, then atoms, and then molecules, stars, plants, cells, and organisms much later. Gradually less complex things came together and became the constituents of more complex things that apparently had very different properties than their less complex parts.

The new mechanistic philosophy offers important insights into the non-reductive behavior of complex systems that could explain why we find a plurality of ontologies in the sciences. But such a vertical approach also struggles with some well-known metaphysical problems of inertness or downward causation that stem from the introduction of a hierarchy of mechanisms. According to this perspective, high-level mechanisms emerge from low-level mechanisms. In order to avoid these conceptual problems, but still being able to uphold a legitimate use of multiple conceptual frameworks in science, I suggest a different perspective. Rather than considering the properties of a system to be either reducible to the intrinsic properties of its parts or emergent properties that supervene on those properties, it is metaphysically simpler and less consequential to think of every system (including the parts) as possessing certain dispositions to interact with that part of a larger system that is considered to be its environment. In this horizontal perspective "emergent" behaviour can be regarded as an extrinsic property, whereas in a vertical perspective this behaviour supervenes on the intrinsic mechanisms of the system. The phenomenon, which the vertical approach considers to be an "emergent" property of the mechanism of the whole system, should – according to this horizontal approach – be regarded as an extrinsic property that results from the interaction between a system and its environment.

According to the horizontal perspective, any self-contained conglomerate of entities that shows some organization and confinement in virtue of certain surface properties or distinguishable borders to the surrounding may count as a system, and what is spatially outside the conglomerate may count as the system's environment. Through interaction with its environment the system becomes part of a larger system that also contains its environment. A system can be a nerve cell, a group of nerve cells, a region of the brain, the whole brain, a starling, a flock of starlings, a habitat, an ecological system, the biosphere, etc. But it is also clear that what the scientist selects as the system, and therefore what is considered to be the environment, is partly context-dependent in the sense that the scientist picks some entity as the analytic unity he wishes to understand.

The explanation of a system's causal behavior in relation to its environment is not reducible to an explanation in terms of the system's own internal *causal mechanisms*. Therefore, I distinguish between internal causal mechanisms and external causal interactions. The internal causal mechanisms only give the system a capacity or disposition to interact in certain ways; whereas the manifestation of that disposition requires that the system actually produces some external responses after having received causal inputs from the environment. Keeping that in mind I hope to have demonstrated that the alternative model of understanding the behaviour of complexity contains some explanatory virtues and fewer metaphysical caveats, and that it can be used to understand why so-called "emergent" properties from atoms to galaxies and from cells to human beings are a consequence of using different conceptual frameworks and descriptions.

Acknowledgement I want to thank Brigitte Falkenburg and Stuart Glennan for insightful comments to an earlier version of this paper.

References

Armstrong, David. 1968. *A Materialist Theory of Mind*. London: Routledge and Kegan Paul.

Craver, Carl. 2007. *Explaining the Brain*. Oxford: Oxford University Press.

Faye, Jan. 2014. What Counts as Causation in Physics and Biology. In *New Directions in the Philosophy of Science*, ed. Maria Carla Gavalotti et al., 173–189. Heidelberg: Springer.

Glennan, Stuart. 2010. Mechanisms, Causes, and the Layered Model of the World. *Philosophy and Phenomenological Research* 81: 362–381.

Kim, Jaegwon. 1999. Making Sense of Emergence. *Philosophical Studies* 95: 3–36.

Lewis, David. 1966. An Argument for the Identity Theory. *The Journal of Philosophy* 63 (1): 17–25.

Machamer, Peter, Lindley Darden, and Carl F. Craver. 2000. Thinking About Mechanism. *Philosophy of Science* 67 (1): 1–25.

Mitchell, Sandra. 2009. *Unsimple Truths. Science, Complexity and Policy*. Chicago: Chicago University Press.

Oppenheim, Paul, and Hilary Putnam. 1958. Unity of Science as a Working Hypothesis. In *Minnesota Studies in the Philosophy of Science*, ed. H. Feigl, M. Scriven, and G. Maxwell, vol. II, 6–36. Minneapolis: Minnesota University Press.

Chapter 10
Crossing Boundaries: Why Physics Can Help Understand Economics

Meinard Kuhlmann

Abstract Socio-economic systems can often successfully be treated as complex systems in the statistical physics sense. This means that the complexity resides in the emerging dynamical behaviour, not in a complicated composition. In order to understand why physics can help to understand socio-economic phenomena with complex behaviour in this sense, I argue that it is necessary to adopt a structural perspective. Accordingly, one has to modify the notion of mechanistic explanations, partly by broadening it. One crucial tool for finding mechanistic explanations in such a structural sense, are minimal models, i.e. models that abstract from micro details in a drastic way. I will show why mechanistic explanations using minimal models for complex dynamical behaviour naturally lead to a structural notion of mechanisms. Among other things I will deal with two diverging views. One of these is that the application of physics-based models to socio-economic phenomena is (mostly) unwarranted to begin with. The other diverging view is that, while the use is in fact warranted, the resulting explanation of socio-economic phenomena is either non-causal or, if causal, not in a mechanistic sense.

10.1 Introduction

Socio-economic systems can often successfully be treated as complex systems in the statistical physics sense. This means that the complexity resides in the emerging dynamical behaviour, not in a complicated composition.[1] In order to understand why physics can help to understand socio-economic phenomena with complex behaviour in this sense, I argue that it is necessary to adopt a structural perspective. Accordingly, one has to modify the notion of mechanistic explanations, partly by broadening it. One crucial tool for finding mechanistic explanations in such a

[1] For a more detailed discussion of this difference, see Kuhlmann (2011).

M. Kuhlmann (✉)
Philosophisches Seminar, Universität Mainz, Mainz, Germany
e-mail: mkuhlmann@uni-mainz.de

B. Falkenburg, G. Schiemann (eds.), *Mechanistic Explanations in Physics and Beyond*, European Studies in Philosophy of Science 11,
https://doi.org/10.1007/978-3-030-10707-9_10

structural sense, are minimal models, i.e. models that abstract from micro details in a drastic way. I will show why mechanistic explanations using minimal models for complex dynamical behaviour naturally lead to a structural notion of mechanisms. Among other things I will deal with two diverging views. One of these is that the application of physics-based models to socio-economic phenomena is (mostly) unwarranted to begin with.[2] The other diverging view is that, while the use is in fact warranted, the resulting explanation of socio-economic phenomena is either non-causal or, if causal, not in a mechanistic sense.

Since the 1990s there is a distinct research field – "econophysics" – with various activities where models and methods from physics are used in order to analyze and explain socio-economic phenomena.[3] Roughly speaking, econophysics consists of three fields of research, which are very different in their nature. In short, one can characterize these fields by the keywords statistical analysis, model building, and applications. In the following study I will primarily look at the first two research fields. Statistical analyses of given empirical data, in particular about financial markets, make up the most firmly established field of research in econophysics. It consists in the analysis of empirical-statistical regularities of prices, e.g. of stocks, without specifying any underlying mechanisms. One main characteristic of this research is the evaluation of very large data sets. Although this task is far more difficult and subtle than one might first expect I think it is justified to say that it is largely concerned with description. To be more precise, it deals with the description of the explanatory targets of econophysics. It is in this field that the empirical phenomena are isolated that econophysics later seeks to explain. That is, it is concerned with explananda.

In the second field of research, econophysicists try to construct models that reproduce the extracted statistical phenomena. In one sub field econophysicists deal with random process models for price fluctuations, e.g. the random walk-model. These models are phenomenological models because their aim is solely to investigate stochastic processes on the macroscopic level without considering any lower level of interacting agents. Another sub field of research is devoted to the construction of microscopic or agent-based models of financial markets, in particular in order to explain the formation of the prices and the evolution of markets. This part of econophysics is concerned with the explanans-side of explanations. Models are constructed in order to explain the empirical features that were extracted by statistical analysis. I will argue that in order to understand how physics-based microscopic models can explain socio-economic phenomena one has to invoke a structural notion of mechanistic explanations.

Already in Kuhlmann (2014) I use the notion of structural mechanisms but with much less emphasis on the following two questions, which are at the core of the present paper: (i) To which extent is it even legitimate to use physics-based models

[2] Thebault et al. (2018) make this point only for one specific approach but apparently see it as a more general issue. See Gallegati et al. (2006) for a more general discussion.

[3] Mantegna and Stanley (2000) is arguably still the best-known introduction to econophysics.

for explaining socio-economic phenomena? (ii) If it is legitimate, is the type of explanation we get really still mechanistic, or even causal at all? In pursuing these two questions, I will deal with some recent literature that critically discusses these very points.

10.2 The Motivation for Physics-to-Economics Import

10.2.1 What Has Physics Got to Do with Human Behaviour?

Econophysicists contend to be able to explain how crashes and bubbles in financial markets come about. But how can physicists contribute anything to the understanding of phenomena that seem to a high degree dependent on what human beings want, fear or hope for? This scepticism is fully justified as far as the actions of single investors are concerned. Physics has nothing to offer for understanding how individuals behave. However, although stock prices are the cumulative effect of the actions of numerous market participants, the collective behaviour of a system with many individuals who influence and thus interact with each other can involve mechanisms that are largely independent from the intentional states of the single individuals. Moreover, these mechanisms are structurally very similar to mechanisms in complex systems that are studied in statistical physics. Here as well one may ignore many microscopic details and physicists have systematically derived reasons why this neglect of micro details is fully legitimate, namely when it comes to the explanation of certain structural patterns in the higher-level dynamics of systems with a large number of interacting components. And it that seems that one can use the same concepts, models and analytical methods, no matter whether one is dealing with magnets or markets.

Many physicists, economists and laymen think that econophysics is a very peculiar kind of science that should not be taken too seriously. As I will show in the following econophysics is a far more conventional science than it first appears. It is a perfectly natural phenomenon in science that analytical and explanatory strategies, which prove to be enormously successful in one context, are also explored in other contexts. Mostly these different contexts lie within one science, but there are also several examples of common explanatory strategies that transgress the borders of one science. One example is the application of evolutionary conceptions for vastly diverse issues far beyond biology. Science lives on the openness for new methods, and scientific breakthroughs were often achieved by the import of methods from one scientific field to another. Econophysics may or may not prove to be fruitful in the end. But one thing is safe to say: The (partially) new way in which econophysicists address old issues follows the path of normal scientific activities.

The most important misconception about econophysics is the thought that physicists claim to be in the position to explain economic phenomena because these are ultimately a part of the physical world and must therefore be explainable in terms of physics. Although it may indeed be justified to think that social phenomena rest or 'supervene' upon their physical manifestation, this is not at all the reason why econophysics works. The possibility of the fruitful transfer of physics' strategies into the domain of economics rests exclusively on structural similarities, and has nothing to do with physics being the most fundamental science. To be slightly more concrete, physics can help to identify and understand mechanisms that generate crashes and bubbles in financial markets, which are structurally similar to mechanisms in certain physical systems. As we will see this structural similarity rests on certain formal analogies.

In Sect. 2.3 I will argue that vis-à-vis complex systems, such as ferromagnets and financial markets and, it makes sense to uphold ontological reductionism while rejecting methodological reductionism. This is compatible with the claim I just made above: Although social phenomena supervene upon their physical manifestation (an ontological claim), it would not be helpful to reduce the behaviour of social systems to the level of elementary particles (a methodological claim). The reason why physics can successfully be used to explain the behaviour of social systems has to do with some insights in physics about the micro-macro relation in complex systems. Originally, these insights referred to genuinely physical systems but they can readily be extended to social systems. In Sect. 4.2 I will argue that such explanations can best be understood in terms of what I call "structural mechanisms".

Explanations in terms of mechanisms – be they classical mechanisms or structural mechanisms – are necessarily *reductive* in some sense because the core idea of mechanistic explanations is reductive: phenomena are explained by reference to interactions between constituent parts. In classical mechanisms, such as a clockwork or an engine, this means that the behaviour of the compound system is reduced to the activities and interactions of its parts, which are described in detail. In structural mechanisms it is also the micro-macro relation that matters. However, as I will argue below, only certain structural features of its interactive organisation matter, whereas the micro details are largely irrelevant, unlike in classical mechanisms. In this respect I have the same diagnose as Christophe Schinckus (2016): he argues that while econophysics defies *atomistic* reduction we still have a case of reduction, namely *interactive* reduction. The point is this (in my terminology again): explanations in terms of structural mechanisms are reductive in the sense that it is the interactive organisation of the micro constituents that matters. However, only certain structural features of the interaction are important whereas the nature of the micro constituents is otherwise irrelevant.

The ensuing investigation is a philosophical case study about a typical kind of complex systems research. In contrast, the philosophy of science, and even more so the philosophy of physics, is traditionally dominated by the analysis of fundamental theories about the most simple objects of the word. While this is certainly very interesting and important, it misrepresents the significance of research about far more complicated composite systems. This neglect is particularly momentous since complex systems research, for instance, has its own repertoire of analytical methods, concepts and explanatory strategies, which cannot be reduced to fundamental science. With the following study I partly wish to contribute to compensating for that deficiency.

10.2.2 Why Adopt a Complex Systems Perspective?

Many quantities peak around some typical value. For instance, adult men peak around 180 cm and the ratio between the heights of the world's tallest and the shortest man is relatively low. There are many such random variables in nature and they lead to a so-called Gaussian (or 'normal') probability distribution, according to which the probabilities for deviations from the mean value decay exponentially. For instance, in the case of body heights this means that the huge majority has heights around the mean value while there are, due to the exponential decay, very few extremely tall or extremely short people. One important feature of such exponential probability distributions is that they indicate a typical size of the measured quantity, namely for instance within three standard deviations around the mean value. Events outside of this characteristic range are extremely unlikely. Exponential probability distributions of natural processes are so wide-spread that it was assumed for a long time that there is no alternative when it comes to randomly distributed quantities.

However, there are also quantities that vary over a huge range of possible values, like the sizes of cities where the ratio of the largest to the smallest population is a six-digit number. Another famous and old example of a quantity that lacks a typical scale is wealth, or income respectively. In fact, it was Vilfredo Pareto's (1848–1923) investigation of the distribution of wealth and his speculations that initiated a whole debate, which continues to the present day.[4] Pareto made the remarkable discovery that the distribution of wealth is different from other allegedly similar quantities, because very large assets contribute a significant proportion to the sum over all inhabitants. Thus for quantities such as wealth, or income, mean values are of little significance since large values occur far more often than according to the Gaussian distribution, and contribute over-proportionally to the total value (which divided by the number of inhabitants yields the mean value). Therefore, the mean value of wealth or income must not be interpreted as a characteristic value, i.e. it is wrong to understand the mean value as expressing what people typically possess or earn.

Today we can make the exact statement that the distribution of wealth does not display exponential but power-law behaviour. Although basically this observation goes back to Pareto, he did not have the analytical means to formulate his results precisely. Mandelbrot (born 1924) had those means and proposed, in the 1950s, an

[4] See Beinhocker's (2007) book *The Origin of Wealth: Evolution, Complexity, and the Radical Remaking of Economics* for a comprehensive up-to-date survey of the debate, which addresses a wider audience and stresses many aspects that are interesting for my investigation of econophysics. Note that Beinhocker talks about 'complexity economics' in contrast to 'traditional economics'.

analysis in terms of what he calls 'Pareto-Lévy distributions', thereby fleshing out Pareto's basic ideas from 1897. Mandelbrot uses certain stable probability distributions that were first introduced in 1925, without any reference to economics, by the mathematician Paul Lévy, one of Mandelbrot's teachers in Paris between 1945 and 1947. Today, one refers to these probability distributions as 'Lévy stable distributions'.[5] One of Mandelbrot's general intentions was to "draw the economist's attention to the great potential importance of 'stable non-Gaussian' probability distributions" (Mandelbrot 1960: 79) for various important economic quantities, including income and wealth among others.[6]

Power-law probability distributions, which figure prominently in Mandelbrot's 'stable Paretian model', are also essential for econophysics. And this is the point where my brief survey of the pre-history of econophysics merges naturally into a discussion of complex systems: many features of complex systems are intimately connected with power-law behaviour and the closely related issue of "scale-invariance". As a last historical remark I should mention that Mandelbrot's path-breaking investigations about power-law (or "fat-tailed") probability distributions of price changes as well as about self-similar structures were the starting point of his later work on complex systems, more specifically on chaos theory and fractals. The basis for Mandelbrot's intellectual development was his realisation that the insights I outlined above are equally valid in numerous other contexts far beyond that of financial markets.[7]

The basic reason why the discovery of innocent-looking power laws arouses the special attention of scientists is because power-law behaviour is tantamount to scale-invariance and thus self-similarity. That is, a power law remains invariant, besides some multiplying constant factor, if one rescales its variable by some factor. It can easily be seen that the power-law distribution $P(x) = Cx^{-\alpha}$ is scale-invariant since

$$P(bx) = C(bx)^{-\alpha} = Cb^{-\alpha}x^{-\alpha} = gP(x)$$

where $g = Cb^{-\alpha}$. In words, rescaling the variable x by a factor b leaves the function P invariant besides the multiplying constant factor g. On a so-called *log-log-scale* a power-law function $P(x) = x^{-\alpha}$ becomes a straight line because

$$\log P(x) = \log x^{-\alpha} = -\alpha \log x$$

[5] The classification 'stable' means that the sum of two independent identically distributed random variables of the Lévy type results in a distribution with the same shape. See Bouchaud and Potters (2000: sect. 1.5.3) for details. The Gaussian distribution is a more common example of a stable distribution.

[6] Also, see Mandelbrot (1963).

[7] See Mandelbrot and Hudson (2004) for a very readable introduction to Mandelbrot's contributions from an *ex post* perspective.

That is, plotting the logarithm of $P(x)$ against the logarithm of its argument x gives a straight line with slope $-\alpha$. Although this sounds as if power-law relations should be immediately visible on a log-log-scaling, it can be surprisingly difficult to confirm presumable power-law behaviour, since one can easily be misled to this conclusion (see Newman 2005).

Complex systems scientists make every effort to discover power-law and therefore scale-invariant relations because it can have far-reaching implications for the behaviour of the system under investigation. Most importantly, under certain conditions, most microscopic details become irrelevant for the dynamics of the system on the macroscopic level. As we will see below for so-called critical phenomena in statistical physics, the occurrence of scale-invariance and hence self-similarity is the deeper reason why diverse systems can exhibit very similar or even identical behaviour, a fact that physicists call 'universal behaviour'. 'Universality' in this sense can be explained via the method of renormalisation involving iterative course graining, which in turn would not be possible without self-similarity. Thus there is a direct road from power-law behaviour, scale-invariance and self-similarity to understanding why certain universal structural mechanisms can account for phenomena in physics as well as in economics. More specifically, power-law behaviour allows applying 'scaling methods', which were first devised in physics, in very different other contexts such as economics.

10.2.3 So What Do Magnets and Markets Have in Common?

Statistical mechanics relates the behaviour of macroscopic systems to the dynamics of their microscopic components or degrees of freedom. This very general description illuminates immediately how physics may be relevant to economics. In both cases one is dealing with, sometimes surprising, statistical features on the macro level and one seeks an understanding in terms of entities on the micro level. Moreover, while the micro components are the basis for explanations, many details about these components don't matter. Rather, at stake is the relation between macroscopic phenomena on the one side and the interaction of a huge number of components on the other side. Since it is the micro-macro relation itself that matters it is conceivable why certain explanatory strategies and models may not depend crucially on whether one is dealing with molecules or with economic agents. One main question is how aggregates, be it magnets or markets, can develop well-defined macroscopic patterns of organization through the undirected interaction of their elements. Figuratively speaking, why does an aggregate move collectively in one direction in the absence of anyone who is 'telling' its components what to do?

One might presume that this is no more of a mystery than the fact that a stone on the tip of a mountain will sooner or later fall 'collectively' to one side or the other. However, due to the solid nature of a stone its components 'have no choice' but to fall into one direction, which is determined by the law of gravitation, the geometry of the arrangement and small symmetry breaking perturbations in the initial and boundary conditions. But dipoles in a ferromagnet as well as traders in a financial market have a choice, in a certain sense at least. While the behaviour of our stone is easily predictable from the behaviour of any one of its components, this is radically different for both ferromagnets and financial markets. Here, knowledge about the behaviour of single components does not lead to any helpful expectations about the behaviour of an aggregate of these components in mutual interaction. One might conclude that this indicates that there are additional irreducible laws that govern certain higher-level systems. Nevertheless, most physicists as well as economists disagree, since usually both are, consciously or not, ontological (although not methodological) reductionists. I share this point of view and propose the following account. Collective behaviour of aggregates is a matter of highly non-trivial dynamical patterns in the simultaneous as well as successive interactions of large numbers of constituent parts. However, such behaviour, which is often denoted as 'emergent', calls for new explanatory strategies and not for extensions of our ontology. Statistical mechanics can boast with a fine repertoire of models, (structural) mechanisms and analytical methods and it is these 'tools' that can be helpful in economics.

Again, one might object. In contrast to molecules, traders can see stock charts and read newspapers, so that they are not just locally influenced. However, in a sufficiently general framework structurally the same can be said about the little dipoles in a ferromagnet. They do feel the magnetisation of the whole magnet. And just as this magnetisation can, under particular circumstances, result from their own interaction, the most interesting features in the behaviour of stocks often emerge endogenously and not due to exogenous factors such as particular economic or political developments.

10.3 An Example: The Spin Model of Financial Markets

The spin model of financial markets presented in Chowdhury and Stauffer (1999) is an early proposal that was used, refined or modified or was an important inspiration in many later approaches.[8] The Chowdhury-Stauffer model rests on the Ising model, which was invented in order to study ferromagnetism, but which may also be readily applied to fluids. Ferromagnetism refers to the mechanism by which certain materials (e.g. iron) become permanent magnets when they undergo a phase transition to their ferromagnetic phase. It arises when a large number of atomic spins align such that their respective magnetic moments all point in the same direction, leading to a net magnetic moment on a macroscopic scale and thus long-range ordering. The simplest theoretical explanation of ferromagnetism can be given with the Ising model.[9] In the well-understood two-dimensional Ising model we have a

[8] This model itself builds upon the simpler and less detailed stock-market model by Bouchaud and Cont (1998).

[9] The model itself was put forward by Ernst Ising (1925). Only in 1944 Lars Onsager famously supplied an exact computation of the thermodynamical properties of the two-dimensional Ising model at the critical point.

two-dimensional lattice of atomic spins, where each node of the lattice can be in one of two different states (e.g. spin up or spin down). Each node influences only the state of its nearest neighbours.

Now, Chowdhury's and Stauffer's idea is to apply the Ising model to financial markets with arrays of traders that may buy or sell just as spins in a ferromagnet can be up or down. In order to make this model work, they draw the following analogies:

Ferromagnet			Financial market
Atomic spins	$i, j, \ldots$	→	Market participants
Spin **up** or spin **down**:	$S_i = \pm 1$	→	Traders' decision to **buy** or **sell**
Strength of interaction between neighbouring spins *i* and *j* with aligning tendency	J_{ij}	→	Herd behaviour
Ground energy	$E_0(t)$	→	Current price
External field	$B_i(t)$	→	Incoming news
Temperature	T	→	Economic situation

With this model Chowdhury and Stauffer (1999: 477) claim to be able to "demonstrate the phenomena of natural and artificially created bubbles and subsequent crashes as well as the occurrence of "fat tails" in the distributions of stock price variations." The fictitious "temperature" is meant to model random influences of the economic situation. The crucial point is that critical phase transitions are seen in parallel with crashes: in both cases we have a drastic change of the state of the entire system without a drastic change in the external conditions. Before I continue with my main question, let me mention that there are various further approaches that use the Ising model to establish an analogy between financial market crashes and critical phase transitions.[10]

10.4 Explaining Complex Dynamics by Structural Mechanisms

10.4.1 Underpinning Analogical Reasoning

Much of econophysics consists in analogical reasoning. Analogies can be an enormous help in calculations. Moreover, they can be of great heuristic value.[11] One of the most important analogies econophysics is based upon is the one between financial market crashes and critical phase transitions. One core issue in judging the

[10] Among them some take Bouchaud and Cont (1998) as their starting point, such as Johansen et al. (2000). Others, like Bornholdt (2001), build upon Chowdhury and Stauffer (1999) – and thereby ultimately upon Bouchaud and Cont (1998), too.

[11] See Hesse (1963) and Bartha (2010) for two extensive philosophical studies on analogies.

legitimacy of this analogy is the role that the renormalization group analysis plays in these two cases. Jovanovic and Schinckus (2016: 58), for instance, argue in their recent book on "Emerging Dialogue" between econophysics and financial economics that "[t]he combination of the renormalization theory and the Ising model offers statistical physicists a unified mathematical framework that can **analogically** be used for the study of phenomena characterized by a large number of interacting microcomponents" (my emphasis in bold face, MK).

The renormalization group method is an ingenious way for iteratively filtering out irrelevant micro details.[12] It was introduced in order to explain the astounding occurrence of universality, i.e. the fact that vastly different materials such as solids and liquids form a "universality class", which means that they can have (almost) the same critical exponent that characterizes their behaviour in phase transitions. Moreover, one can show that the Ising model lies in the same universality class as ferromagnets and fluids. This fact justifies using the Ising model with its highly idealized assumptions in order to explain critical phase transitions in those two systems. The renormalization group analysis shows us that the idealized assumptions of the Ising model do not harm, because the assumptions about the micro details that may seem problematic, in particular the unrealistic lattice structure, get washed out anyhow. Thus it can be shown that the idealizations of the Ising model fulfil their purpose: They simplify the calculations without disturbing the result in which we are interested.

Now, the crucial question is whether it is justified to apply the Ising model for the analysis of financial markets, in a similar way as it is justified to apply it to diverse materials in the same universality class. Unfortunately, it seems that we are here dealing with a case where analogical reasoning fails because the source and the target domain are disanalogous in at least one crucial aspect: Financial market crashes do not form a universality class, at least not in this sense specified above. Market crashes exhibit quite different "critical" exponents. Thus it does not seem to be justified to treat financial market crashes and critical phase transitions in an analogous fashion. Both Knuuttila and Loettgers (2016) as well as Jhun et al. (2017) address this worry but with diverging conclusions.

Knuuttila and Loettgers (2016: 395) argue that in the "context of its socio-economic applications, one cannot justify the use of the Ising model in the same way as one does in physics", namely by renormalization group methods. Instead, they argue, four elements are needed for such a justification:

(i) The starting point is "some empirically observed patterns" such as (volatility) clustering.
(ii) With the Ising model one has a "well-understood mathematical and computational method to model such phenomena". Seemingly, Knuuttila and Loettgers take it that only these two points are usually in place in econophysics, and, they

[12] See Binney et al. (1992) for a detailed exposition and Batterman (2002), Morrison (2014) as well as Jovanovic and Schinckus (2016: 58) for accessible introductions together with philosophical/conceptual discussions.

argue, this is not enough to justify using the Ising for socio-economic issues. They claim that one needs two further ingredients, which are not, or at least not sufficiently, supplied in econophysics.[13]

(iii) The third ingredient is "a very general conceptual idea of the kind of structure or interaction that the model exhibits". Unfortunately, it remains a bit vague, what they mean exactly. They mention clustering and phase transitions as examples, but in my view, this still does not make it entirely clear. On the face of it, "kind of structure or interaction" seems to refer to two different issues, namely, on the one hand, *macroscopic* structure, and, on the other hand, *microscopic* interaction. Moreover, it is a bit surprising that clustering is given as an example once again. What is not rather an example for an "empirically observed pattern", i.e. for ingredient (i) of a justification? Moreover, if clustering is a convincing example for fulfilling requirement (i), why is it no longer convincing when it comes to requirement (iii) – assuming that it is an example for (iii) to begin with? In any case, the main point of Knuuttila and Loettgers is that it is not enough to take some observed economic pattern and use the Ising model to explain it because one knows from physics contexts that the Ising model can be used to explain observed (physical) phenomena with similar statistical characteristics as in the economic cases.

(iv) Knuuttila's and Loettgers' last requirement is "a new interpretation in social psychological terms that gives initial plausibility to the model". In the end Knuuttila and Loettgers (2016: 398) conclude that "[w]hen models [such as the Ising model] are transferred from one discipline to another, they may lose large parts of their original theoretical and mathematical grounding, thereby turning into thin analogue models."

Thus, Knuuttila and Loettgers reach a negative evaluation concerning the worry that the use of phase transition models (such as the Ising model) is no longer underpinned by the renormalization group method when they are applied to socio-economic phenomena. This evaluation is not inevitable, however. Jhun et al. (2017) (henceforth "JPW") counter the worry in the following way: It is true that financial market crashes do not constitute a universality class (in the sense specified above) so that the standard renormalization group approach to critical phase transitions is inapplicable. However, JPW argue, the explanatory goal is different in these two cases and so is the adoption of the renormalization group technique. In the case of critical phase transitions the goal is to explain why vastly different materials have (almost) the same critical exponents. This reasoning cannot be transferred to the analysis of financial market crashes, however, since these do not constitute a universality class, in the sense of the renormalization group technique. Does this mean the adoption of the Ising model loses its justification when it comes to explaining financial market crashes?

[13] Note that Knuuttila and Loettgers look at socio-economic applications in general and not specifically at econophysics. In fact, they don't even mention the term 'econophysics'. However, they study the work of some of the founders of econophysics, such as Stanley and Stauffer, and, in any case, their considerations seem equally relevant for econophysics.

JPW (p. 14) argue that

> "if there is universal behavior in stock market crashes, this is not the kind of universal behavior that can be explained via RG methods alone. [...] one cannot apply the same reasoning as in physics to argue that large-scale market behavior near transition points (i.e., crashes) is independent of the microscopic details of market dynamics. It thus seems that insofar as the JLS[14] model is successful, it must function differently. [...] [It] relies on a subtle interplay between microscopic and macroscopic considerations [...]."

JPW go on to argue that what one needs in addition is (discrete) scale invariance. More specifically they claim (JPW: 16f) that

> "it is the inference [...] to discrete scale invariance of an underlying network structure (or, more generally, from power laws of any kind to scale invariance) that forms the explanatory core of the JLS model. In more detail, what we find here is an explanation of (endogenous) market crashes as arising from the structure of the network of traders at the time the crash occurs. Markets crash in the absence of any external, coordinating event because the network of traders can spontaneously evolve into states that are (discretely) scale invariant, i.e., which have long correlation lengths, so that small, essentially arbitrary perturbations, can propagate rapidly across scales".

Up to this point I perfectly agree to the analysis of JPW. However, now they take a turn I think is unwarranted. They argue that

> *"this explanation, as we understand it, is causal, in the sense of Woodward's interventionist account of causation [...]: if the network of agents participating in a market approaches a (discretely) scale-invariant state [...], then (it is likely that) a crash will occur. In other words, the model says that crashes occur in many different systems precisely when their (coarse-grained) dynamics become approximately discretely scale invariant. And so, it is the emergence of discrete scale invariance (or, perhaps, scale invariance more generally) that should be identified as the proper cause of the crash"* (JPW: 16f).

> *"[It] is (discrete) scale invariance that causes market crashes—or, to put it in more evocative terms, it is herding at all scales that causes market crashes"* (JPW: 20).

I think it is categorically inappropriate to say that scale invariance is a cause. Herding, in contrast, is not merely the same "in more evocative terms": herding can be a cause. Scale invariance is the mathematical signature that points to causal processes. Scale invariance arises from how the system's parts interact with each other. Scale invariance supervenes upon the interactive organization of the system; it is not itself a causal agent. More specifically, scale invariance is not a variable that one could manipulate so that it fits Woodward's (2002) interventionist notion of a cause. However, this would have to be the case, since it is this very account that JPW claim to capture the sense in which market crashes are caused. Roughly, the core idea of Woodward's is that variable A has a causal effect on variable B if and only if it is possible to manipulate only A in a way (called "intervention") that its changes are accompanied by corresponding changes in variable B. However, scale invariance is

[14] JPW are primarily concerned with one particular example, namely the financial market model of Johansen et al. (2000) (henceforth "JLS"). It is similar to the example I introduced above, at least for the purposes of the current discussion. They focus on the question to which extent this model does in fact treat crashes and critical phase transitions as analogous.

not such a variable. You cannot directly manipulate scale invariance. One can only manipulate certain features of the interactive organization of the system so that scale invariance arises.

In the light of these considerations it seems unwarranted that JPW (2017: 21) continue

> "[…] we do not claim that crashes are being explained, here, by appeal to particular details concerning interactions between individual agents. In this sense, it is not a "causal-mechanical" or "mechanistic" explanation […]. Indeed, the model is not committed to any particular network model at the microscale, just a class of models that exhibit discrete scale invariance."

In the following section I will lay out a reading of the econophysics explanation of market crashes that is exactly this: mechanistic. More precisely, it is an account of the explanatory accomplishments of econophysics in terms of structural mechanisms. The following quotation shows that JPW, somewhat surprisingly, seem to see the same point but make less out of it than I propose to do. JPW (2017: 21f) concede that

> *"[…] the JLS model […] refers to the* ***micro-constituents*** *of the market. It is in this sense that we take the explanation to be reductive: it explains a phenomenon by appealing to relations between the parts of a system – in this case,* ***interactions*** *that occur between agents in a network. […] But it does not follow that the model supposes an atomistic conception of the economy, i.e. it does not determine the law governing the behavior of any arbitrary agent. Given some behavioral assumptions, it does constrain the kinds of* ***structures*** *they might reside in. In this case: hierarchical* ***structures*** *that (sometimes) exhibit discrete scale invariance. This does not require any particular arrangement of individuals because those particular details are in some sense irrelevant; what does matter are these* ***structural details****"* (my emphasis in bold face, MK).

I take this to be the core point of the mechanistic account of explanation: It is the interactive organization (between the parts of the mechanism) that does the explaining. And sometimes it is not all the details of the interactive organization that matter but just certain structural details of it.

10.4.2 Structural Mechanisms

As we saw above, universality and renormalization are currently vividly discussed in philosophy. Much of the discussion turns on the claim by Batterman in particular that "minimal models" involving "infinite idealizations" in terms of the renormalization group constitute a genuine new type of explanation. More specifically, it is claimed to be a non-causal explanation because the micro details do not matter.[15] In the following I want to argue that the use of minimal models in this sense does not necessarily result in non-causal explanations.

[15] See Batterman (2002) and Batterman and Rice (2014).

Concerning econophysics it is particularly hard to evaluate the applicability of causal accounts of explanation. Major parts of econophysics are concerned with the statistical analysis of financial data and it is difficult in general to identify causes in statistical theories, since nothing is said about particular causes. Rather it is said which kinds of configurations of interacting constituents lead to which statistical properties on the macro level. This problem for causal accounts of explanation applies generally to analyses of higher-level systems and in particular complex systems.[16] One characteristic feature of complex systems consists in the fact that compounds of materially very different micro constituents can exhibit the same macro behaviour. Since in many contexts only the macro behaviour is of interest, the knowledge of micro-details does not contribute anything to what one wants to understand. This fact is a problem for causal accounts of explanation since here the core of explanations is seen in the specification of causes that lead to particular events. Since it seems that only material constituents or fields can be causally efficacious but not general patterns it is hard to see how causal accounts of explanation can shed any light on the explanatory force of analyses of complex systems.

Although prima facie the previous argument sounds convincing, in fact it rests completely on one particular notion of causation. According to this notion, causation consists in continuous spatio-temporal processes, e.g. with a transfer of energy and momentum, on the level of individual constituents of a complex system. And this conception of causation indeed seems not suited to understand causal factors in agent-based models. However, there are strong competitors to this notion of causation in terms of physical processes. In the last 15 years arguably the most important, and definitely most discussed, rival account is the interventionist theory of causation by Woodward (2002) we encountered above already. Yet another related proposal is the mechanist account of causation by Glennan (2017, chap. 5), which is a singularist account. Both accounts, Woodward's and Glennan's, are non-reductive, which is one reason why both have been challenged as genuine accounts of causation. However, for my purposes, what matters is not whether we have a mechanistic account of causation or not. The crucial thing is that we have a *mechanistic* account of causal *explanation*, because this account is particularly apt to understand the generation of higher-level phenomena.

So what exactly is a mechanism? Above I already mentioned clockworks and engines as paradigm mechanisms. In the recent philosophical discussions, biological mechanisms in cells and the brain have also received a lot of attention, e.g. the circadian rhythm. After various attempts for defining the notion of a mechanism, most notably by Machamer et al. (2000) and Glennan (1996, 2002), there have been some proposals for a consensus characterisation, such as:

> A mechanism for a phenomenon consists of entities and activities organized in such a way that they are responsible for the phenomenon.
>
> (Illari and Williamson 2012, p. 120)

[16] See Woodward (2002: sect. 4.3) for a very illuminative discussion.

and

> "A mechanism for a phenomenon consists of entities (or parts) whose activities and interactions are organized so as to be responsible for the phenomenon."
>
> (Glennan 2017)

On this basis I see three essential elements for a mechanistic explanation of the behaviour of a given system:

(1) Specify the phenomenon for which an explanation is sought,
(2) identify the relevant parts of the system and their interactive properties that allow you to
(3) show how their interactive organization is responsible for the phenomenon to be explained.

While extensive statistical analyses of financial data are indispensable for all other enquiries in econophysics and sufficient for some practical purposes, I think it is justified to say that no pretence to explanation is made. It is described and mathematically analysed what one observes but no story is told about how these observed facts are produced. Of all accounts of explanation the causal-mechanical model emphasises most that it is necessary to say what it is in the world that brings about the explanandum event. This is exactly the aim of the other main part of econophysics where, in a clear-cut division of labour, 'stories' are supplied that may explain the observed statistical features on the basis of microscopic models of financial markets. I want to argue that physics-based microscopic models of financial markets do specify a causal mechanism on the level of traders, which leads to the statistical effects for which an explanation is sought.

In Kuhlmann (2014) I claim that phase transitions in ferromagnets and financial markets can be studied in a common framework because the same structural mechanisms can be invoked in both cases. I argue that in order to identify a common mechanism across radically diverse complex systems one needs to focus on certain structural features. I propose to distinguish two different classes of structures in structural mechanisms, namely (i) structural start and boundary conditions, and (ii) emerging dynamical structures. If one has identified a structural mechanism, then one knows that a certain set of structural start and boundary conditions (i) is essential for producing certain dynamical structures (ii). What one needs in order to corroborate that one has really found a mechanism and not just an artefact is the fulfilment of a certain robustness condition.

Structural start and boundary conditions – i.e. the first class of structures in structural mechanisms – concern aspects such as connectivity, dimensionality, topology and certain symmetry properties. Connectivity is the most important structural aspect. The crucial issue in complex systems is the interaction between the system's parts, and neither their detailed behaviour nor their minute spatio-temporal organization in the whole system. What really matters is the dynamical interactive organization of a complex system, and even there only certain structural aspects. A conventional mechanistic explanation shows how the often sequential interactions of the different parts, which fulfil specific functions, produce a certain behaviour. In complex systems it is usually impossible or at least not helpful to distinguish parts with different functions that play specific stable roles in the mechanism. Mostly all parts have identical properties and behaviour. What is essential instead are the structural features of their interaction. Moreover, the parts of a complex system usually all interact simultaneously. In contrast to conventional mechanisms, all parts are governed by the same behavioural rules and no external force tells them what to do. The relevant start and boundary conditions for structural mechanisms in complex systems do not describe a specific configuration that already allows imagining what will happen if we let the system run. This is a crucial difference to conventional mechanisms. In complex systems, for interesting things to happen, it suffices to have, in a sense, an amorphous set-up with very general structural properties that apply to the whole system.

However, structural start and boundary conditions (i.e. the first class of structures in structural mechanisms) is not enough. One also needs a second class of structures in structural mechanisms, namely ones concerning the emerging dynamical structures. This refers to the fact that non-trivial long-range effects, for instance phase transitions, arise dynamically purely on the basis of short-range interactions. The entire system behaves as if there were some external coordination, while in fact there is none. In fact, I argue that this is one of the deep ideas behind the notion of mechanisms: Once it is set up in the appropriate way it runs largely by itself without the need for any further coordination. But there is one pivotal issue that distinguishes conventional mechanisms from structural mechanisms in robust complex behaviour – which is also the reason for the term "self-organization": In complex systems the "organization" that is crucial for the system behaviour is not already present in the initial set-up but only arises through the dynamics of the system, namely by the interaction of its parts.

It is only possible to say that there is a common mechanism in diverse systems such as ferromagnets and financial markets if one stays on the structural level. For example, one doesn't want to claim that market traders actually sit on a grid and only interact with their spatially nearest neighbours. Rather, the crucial point is more abstract or structural: Large changes, be it phase transitions or financial-market crashes, and other related phenomena, can arise purely from the mutual interactions of the systems' parts without any external coordination. One task of complex systems theories is to identify the underlying structural mechanism.

The fact that the structural mechanisms, which I have argued to be essential for an understanding of financial markets, are situated on an intermediate level, necessitates pragmatic reasoning for their identification. Explanations can thus also have aspects that are relative to our interests. We are interested in the dynamics of the macrostate so that the concentration on the macrostate and the neglect of micro-details is at least partly driven by our interests. And these interests are particularly important in the context of financial markets because applications are the ultimate goal of econophysics. Moreover, pragmatic considerations are particularly important for complex systems due to the fact that they are hard to handle analytically.

The covering law model of explanation as well as the unificationist, the causal (mechanical) and the pragmatic account all have something to offer in describing how microscopic models of financial markets explain. However, they also all have their shortcomings. Three reactions recommend themselves. The first is to deny that there is a single theory of scientific explanation that covers most or even all cases of explanations that are generally seen as successful. The second possibility is to construct an eclectic combination. The third option, which I favour, is to revise or extend the most promising candidate. I think this candidate is the causal-mechanical model of explanation, which needs to be extended to include, first, an analysis of universal behaviour that sets the stage, and second, a structural notion of mechanisms. Although for these mechanisms, in contrast to biological mechanisms, the micro details of the material constituents to large degree make no differences to the macro outcomes, I conclude that they are still legitimately called mechanisms since they describe, in a certain sense, robust structural input-output relations. Eventually, although I think that a revised causal-mechanical account of explanation is best suited to represent how econophysics explains, the remaining three aspects of explanation, namely subsumption, unification and pragmatics, should still not be neglected, since they round up and complete the picture.

10.4.3 The Role of Models in Interdisciplinary Complex Systems Approaches

The role of models is crucial for understanding how econophysics explains.[17] The two most important issues in this respect are idealisation and the discovery of mechanisms. Idealisation matters because the models employed in econophysics deliberately simplify things so that they are, strictly speaking, false. Nevertheless, despite of their falsity, they are an indispensable and highly efficient tool for understanding the real world. Obviously, this calls for a clarification. The other closely connected aspect concerns mechanisms. As I argued above, one particularly attractive account for the causal aspect of explanation rests on the notion of generating mechanisms. However, while the mechanism in a classical watch is immediately visible, mechanisms in complex systems, such as the ones studied in econophysics, are far less accessible. As it turns out, one needs a fairly complicated strategy of modelling in order to identify and understand mechanisms in complex systems. Thus models play a crucial role in determining causal relations, which constitute the most important aspect of explanations in econophysics.

One of the most important aspects where econophysics resembles economics more than physics is the predominance of model building in the absence of uncontroversial underlying theories. Thus although one may distinguish 'observational econophysics' from 'theoretical econophysics' (Roehner 2002: xii) this does not mean that econophysics attempts to formulate, in a classical sense, theories for the

---

[17] See Weisberg (2013) for a comprehensive discussion of scientific models, Gibbard and Varian (1978) for a classical paper specifically on economic models, and Wimsatt (1987) on the function of "false models," which is particularly relevant to complex systems.

behaviour of financial markets, for instance. Rather, the objective of what one may call 'theoretical econophysics' is to explore the underlying mechanisms that explain the observed behaviour. Therefore, the distinction of 'observational econophysics' and 'theoretical econophysics' is rooted in the difference between description and explanation. And as it turns out, the main way to identify underlying mechanisms is via modeling the resulting dynamics of a large number of interacting agents in a financial market. Thus the causal explanation of observed economic phenomena in terms of generating mechanisms crucially depends on model building. In what Roehner calls 'observational econophysics' models are also of central importance, although a different type of model, namely phenomenological models.

Econophysics does not try to establish general theories about, e.g., financial markets but is, like economics, mostly rather concerned with model building. At least two reasons come to mind why econophysics is restricted to model building at the expense of the formulation of a theory. The first of these is that econophysics is still in an exploratory stage and yields only preliminary results—not just in the sense as any science is preliminary. If that should be the reason for the predominance of model-building activity in econophysics, then its models would function as instruments for theory construction and exploration in the absence of an established theory. This is arguably the view to which most scientists would subscribe. In the case of econophysics the exploratory aspect of models is certainly strong and not to be denied. Nevertheless, I think there is another important reason for the predominance of models in econophysics. One of my main theses about econophysics is that its models describe mechanisms that function prominently in causal-mechanical explanations of the target phenomena of econophysics. I think this way of reasoning fits well to the analysis in Morrison (1999) and in Morgan (1999) according to which models can be relatively autonomous from theory and can be instruments for the representation of the world, i.e. not for the representation of the theory.

In econophysics the notion of models matters in a number of important ways and brings up various questions. Arguably the best examples of successful explanations in econophysics involve microscopic models of financial markets. These are highly idealized agent-based models with complex behaviour. This raises a number of questions. What are the ontological, the epistemological status of these models, i.e. what kinds of entities are these models and what is their function in the scientific discovery and understanding? Moreover, what is the semantics of models, i.e. what and how do models represent? Closely related to models is the issue of idealisation. How can highly idealized, and therefore strictly speaking false models help to understand or even represent reality? Moreover, is the origin of the above-mentioned models in physics or in economics? And connected to that question, is the relevant context for econophysics economic or physical models or possibly a third, more abstract issue?

Often the models that are used in econophysics come from economics or other social sciences, such as the plethora of agent-based and game-theoretic models. Moreover, the important random-walk model was also developed in an economic context, namely Bachelier's famous analysis of speculative prices in stock markets. But econophysics also deals with models that stem from physics, such as the Ising model or the nearest neighbour models. However, arguably the most important contribution from physics consists in analysing the dynamical behaviour of economic models and in identifying underlying mechanisms, which are often well understood in physical contexts.

From a philosophical perspective it is puzzling that the explanatory strategies in statistical physics and economics are so similar. The surprise about this similarity is partly due to the fact that in physics fundamental theories are available while this is not the case for large areas of economics, e.g. concerning financial markets. Nevertheless, in both statistical physics and econophysics models are used that are to a certain degree independent from an underlying theory. But the reason for the partial independence of models from theory seems to be a different one in physics and economics. While in statistical physics fundamental theories are often not needed since various micro details are irrelevant for many questions, widely accepted fundamental theories are simply inexistent for many economic issues. The inexistence of fundamental theories in economics need not imply that models only act as substitutes or even makeshifts. It is the very similarity between phenomena in statistical physics, in particular solid state physics, and economics that justifies neglecting micro details, which would have to be described by fundamental economic theories. In view of this similarity the use of models in econophysics might not be due to the unavailability of relevant theories and the preliminary status of research but is rather conditioned by certain structural features that systems in both statistical physics and econophysics share.

Many publications on statistical mechanics contain general reflections on the role of models as well as systematic treatments of used models. A large proportion of the models that are used in statistical mechanics are so-called toy models, which are not expected to be realistic from the outset. It is important to realize that toy models are usually not meant to help newcomers or in order to avoid tedious work with more complicated models. Objectives are rather finding out which features are realistic, understanding other models and testing where exactly toy models fail, which can be particularly instructive.

10.4.4 Models and Mechanisms

In non-law-based accounts of explanation, such as Woodward's (2002) causal approach, models often play a crucial role.[18] For instance models can figure as devices for discovering causal relations, which are core ingredients of causal explanations. And if one thinks that these causal relations are best understood in terms of mechanisms, then models are tools for finding out about mechanisms, which in turn do the explanatory job. In my view, the identification and understanding of

[18] Cartwright's (1983: ch. 8) so-called 'simulacrum account of explanation', which is also non-law-based, is arguably the approach where models and explanations stand in the closest connection with each other.

mechanisms is in fact the main function of models in econophysics. Moreover, I think this evaluation is in accordance with the role of models in important accounts of mechanisms, such as the ones by Glennan (2017) and by Woodward (2002).

In Glennan's (2017: ch. 3) conception of mechanistic explanations the notion of 'mechanical models' is of great importance. The formulation of a mechanical model has to supply two descriptions. First, it has to describe the behaviour of a mechanism, i.e. the explanandum. The second part of a mechanical model, the explanans, consists in the description of a mechanism that accounts for the respective behaviour. Thus a mechanical model comprises a behavioural description (the explanandum) and a mechanical description (the explanans). The distinction between the description of external behaviour and internal structure, i.e. the workings of the mechanism, corresponds to the difference between "what a system is doing" and "how it is doing it" (Glennan 2002: S347). As an illustration for a mechanistic explanation of a general regularity Glennan discusses input-output mechanisms, such as a word processing program on a computer, in which case the input is the pressing of certain keys on the keyboard while the output are the characters that are produced on the computer screen. The mechanistic explanation of input-output mechanisms consists in showing how the input triggers a chain of interactions between parts of the mechanism.

Mechanisms are not simply very detailed descriptions of what is happening, but their identification is crucial for causal explanations of why things behave the way we observe them. The specification of mechanisms is explanatory because it abstracts from as many details as possible with respect to the explanatory target. Thus simplicity is a crucial characteristic of mechanisms and the best way to identify mechanisms in complex systems is by constructing simple idealized models.[19] The connection of models and mechanisms can be illustrated with respect to multi-agent models of social phenomena, such as the microscopic model of financial markets above: These models are dynamical models, i.e. they describe how aggregate social entities evolve in time. By running a microscopic model of a financial market on a computer one may reproduce the statistical features of financial markets one wants to explain. However, this potentially accidental brute success in a computer simulation is not enough for an explanation. In order to get an explanatory account one has to identify robust dynamical features of the model that describes the production of the observed higher-level patterns. In short, one needs to identify mechanisms.

[19] Batterman (2005) has a similar point, when he argues that highly idealized and oversimplified models can sometimes be better for the explanation of the dominant phenomenon, e.g. a phase transition, than a detailed model in terms of micro-constituents.

10.5 Conclusion

Up to the 20th century, mechanistic thinking was closely related to atomistic conceptions of matter and was opposed to the assumption of hidden powers and entelechies. Today, approaches that stress the significance of mechanisms have a different opponent. Mechanistic models of causal explanation rest on the idea that it is not enough for an explanation of a phenomenon to show that is to be expected on the grounds that its statement is the logical consequence of a statement about some regularity in nature. What is missing is the specification of what it is in the world that brings the phenomenon about. In other words, one needs to identify a mechanism that describes how the phenomenon is generated by the interactive organization of constituent parts.

The mechanistic account of explanation is also attractive when it comes to interdisciplinary approaches such as econophysics, where insights from physics are used to understand socio-economic phenomena. Econophysicists build dynamical models which reproduce certain statistical phenomena that call for an explanation. One of my main claims is that the explanatory models devised in econophysics are only successful if they allow for the identification of generating mechanisms. However, and here comes the problem I have dealt with, the mechanisms that matter in econophysics need to be characterized differently from the more classical ones, such as in clockworks or biological cells. The reason is that econophysics treats financial markets as complex systems, e.g. in terms of multi-agent models. Although the biological objects for which mechanisms are studied are also often called 'complex systems', the complexity involved differs fundamentally from that in complex systems theories such as econophysics. While the complexity of what I prefer to call 'classical mechanisms' consists in their complex compositional structure, proper complex systems exhibit their complexity in complex dynamical features, while the arrangement of their parts is often even particularly simple.

The crucial points of my characterisation of structural mechanisms in complex systems are the following. One needs to distinguish a higher level of quantities that refer to the whole composite system and a lower level of (mostly a large number of) interacting parts. Mechanisms are always relative to some higher-level type of pattern one wants to explain. Moreover, only those dynamical features qualify as mechanisms that come about purely by the mutual (and usually non-linear) interaction of the system's constituents and not by any external coordinating influence. Eventually, the identification of a mechanism is only successful if the type of pattern one seeks to explain arises in a robust way, i.e. if the pattern remains qualitatively invariant under perturbations of the system's parameters, within a reasonable range of values.

Along the way I have defended my approach against an alternative reading of explanations in econophysics. According to this alternative view, the mechanistic account of causal explanation is ruled out because atomistic details do not matter while the proper cause of market crashes is claimed to be scale invariance. I argue that scale invariance does not qualify as a cause but only arises from the interactions of the system's parts. Scale invariance is a signature of what is going on and not itself a cause. Remarkably, in complex systems only certain structural features of the interactive organization matter to give rise to market crashes. However, this does not speak against mechanistic explanations *per se* but only against an understanding where all micro details matter. In interdisciplinary complex systems it is still the interactive organization of the system's parts that matters, even if only certain structural features of it.

Literature

Bartha, P. 2010. *By Parallel Reasoning: The Construction and Evaluation of Analogical Arguments*. Oxford: Oxford University Press.
Batterman, R.W. 2002. *The Devil in the Details*. Oxford: Oxford University Press.
———. 2005. Critical phenomena and breaking drops: Infinite idealizations in physics. *Studies in History and Philosophy of Modern Physics* 36: 225–244.
Batterman, R.W., and C.C. Rice. 2014. Minimal Model Explanations. *Philosophy of Science* 81 (3): 349–376.
Beinhocker, E.D. 2007. *The Origin of Wealth: Evolution, Complexity, and the Radical Remaking of Economics*. London: Random House Business Books.
Binney, J.J., N.J. Dowrick, A.J. Fisher, and M.E.J. Newman. 1992. *The Theory of Critical Phenomena: An Introduction to the Renormalization Group*. Oxford: Oxford University Press.
Bornholdt, S. 2001. Expectation bubbles in a Spin Model of Markets: Intermittency from Frustration Across Scales. *International Journal of Modern Physics C* 12: 667–674.
Bouchaud, J.-P., and R. Cont. 1998. A Langevin Approach to Stock Market Fluctuations and Crashes. *European Physical Journal B* 6: 543–550.
Bouchaud, J.-P., and M. Potters. 2000. *Theory of Financial Risks: From Statistical Physics to Risk Management*. Cambridge: Cambridge University Press.
Cartwright, N. 1983. *How the Laws of Physics Lie*. Oxford: Clarendon Press.
Chowdhury, D., and D. Stauffer. 1999. A Generalized Spin Model of Financial Markets. *European Physical Journal B* 8: 477–482.
Gallegati, M., S. Keen, T. Lux, and P. Ormerod. 2006. Worrying Trends in Econophysics. *Physica A: Statistical Mechanics and Its Applications* 370: 1–6.
Gibbard, A., and H.R. Varian. 1978. Economic Models. *The Journal of Philosophy* 75: 664–677.
Glennan, S.S. 1996. Mechanisms and the Nature of Causation. *Erkenntnis* 44: 49–71.
———. 2002. Rethinking Mechanistic Explanation. *Philosophy of Science* 69: S342–S353.
Glennan, S. 2017. *The New Mechanical Philosophy*. New York: Oxford University Press.
Hesse, M. 1963. *Models and Analogies in Science*. London: Sheed and Ward.
Illari, P.M., and J. Williamson. 2012. What Is a Mechanism? Thinking About Mechanisms Across the Sciences. *European Journal for Philosophy of Science* 2: 119–135.
Ising, E. 1925. Beiträge zur Theorie des Ferromagnetismus. *Zeitschrift für Physik* 31: 253–258.
Jhun, J., Palacios, P., and J. Weatherall. 2017. Market Crashes as Critical Phenomena? Explanation, Idealization, and Universality in Econophysics. *Synthese* (online first).
Johansen, A., O. Ledoit, and D. Sornette. 2000. Crashes as critical points. *International Journal of Theoretical and Applied Finance* 3: 219–255.
Jovanovic, F., and C. Schinckus. 2016. *Econophysics and Financial Economics: An Emerging Dialogue*. New York: Oxford University Press.
Knuuttila, T., and A. Loettgers. 2016. Model Templates Within and Between Disciplines: From Magnets to Gases – And Socio-economic Systems. *European Journal for Philosophy of Science* 6: 377–400.

Kuhlmann, M. 2011. Mechanisms in Dynamically Complex Systems. In *Causality in the Sciences*, ed. P. McKay Illari, F. Russo, and J. Williamson, 880–906. Oxford: Oxford University Press.

———. 2014. Explaining Financial Markets in Terms of Complex Systems. *Philosophy of Science* 81: S1117–S1130.

Machamer, P., L. Darden, and C.F. Craver. 2000. Thinking About Mechanisms. *Philosophy of Science* 67: 1–25.

Mandelbrot, B. 1960. The Pareto-Levy Law and the Distribution of Income. *International Economic Review* 1: 79–106.

———. 1963. The Variation of Certain Speculative Prices. *Journal of Business* 36 (4): 394–419.

Mandelbrot, B., and R.L. Hudson. 2004. *The Misbehavior of Markets: A Fractal View of Risk, Ruin and Reward*. London: Basic Books.

Mantegna, R.N., and H. Eugene Stanley. 2000. *An Introduction to Econophysics: Correlations and Complexity in Finance*. Cambridge: Cambridge University Press.

Morgan, M. 1999. Learning from Models. In *Models as Mediators: Perspectives on Natural and Social Science*, ed. M.S. Morgan and M. Morrison, 347–388. Cambridge: Cambridge University Press.

Morrison, M. 1999. Models as autonomous agents. In *Models as Mediators: Perspectives on Natural and Social Science*, ed. M.S. Morgan and M. Morrison, 38–65. Cambridge: Cambridge University Press.

———. 2014. Complex Systems and Renormalization Group Explanations. *Philosophy of Science* 81: 1144–1156.

Newman, M. E. J. 2005. Power Laws, Pareto Distributions and Zipf's Law. *Contemporary Physics* 46: 323–351, citations from third online version: http://aps.arxiv.org/abs/cond-mat/0412004v3.

Roehner, B.M. 2002. *Patterns of Speculation: A Study in Observational Econophysics*. Cambridge: Cambridge University Press.

Schinckus, C. 2016. 1996-2016: Two Decades of Econophysics: Between Methodological Diversification and Conceptual Coherence, Special Topics issue 'Discussion & Debate: Can Economics be a Physical Science?'. *The European Physical Journal* 225: 3299–3311.

Thebault, K.P.Y., S. Bradley, and A. Reutlinger. 2018. Modelling inequality. *British Journal for the Philosophy of Science* 69: 691–718.

Weisberg, M. 2013. *Simulation and Similarity: Using Models to Understand the World*. New York: Oxford University Press.

Wimsatt, W.C. 1987. False Models as Means to Truer Theories. In *Neutral Models as a Biological Strategy*, ed. M.H. Nitecki and A. Hoffman, 23–55. Oxford: Oxford University Press.

Woodward, J. 2002. *Making Things Happen – A Theory of Causal Explanation*. Oxford: Oxford University Press.

Chapter 11
Realizing Computations

Vincenzo Fano, Pierluigi Graziani, Mirko Tagliaferri, and Gino Tarozzi

Abstract The aim of this paper is to address the question: when does a physical system realize (implement) a certain computation? The most developed account that answers this question is Piccinini's mechanistic account. Our strategy is to start from Piccinini's reflections, emphasizing different aspects of the problem of realization and thus proposing a novel account. Our idea is to propose a new definition of realization that makes the original question more tractable and easier to scrutinize. We show that our definition has some advantages when dealing with classical objections to accounts of computation in physical systems.

The paper is structured in four parts: after the introduction, the first part will introduce mapping accounts of implementation discussing some of their problematic aspects; the second part will present and clarify some prerequisite notions for a definition of realization; the third part will introduce our definition – it will turn out that our definition will identify a specific kind of strategy that Piccinini (2015a, b) calls nomological mapping account; the fourth and final part will be dedicated to analysing the advantages of our definition. Concluding remarks follow.

V. Fano (✉) · P. Graziani · M. Tagliaferri · G. Tarozzi
Department of Pure and Applied Sciences, University of Urbino Carlo Bo, Urbino PU, Italy
e-mail: vincenzo.fano@uniurb.it

B. Falkenburg, G. Schiemann (eds.), *Mechanistic Explanations in Physics and Beyond*, European Studies in Philosophy of Science 11,
https://doi.org/10.1007/978-3-030-10707-9_11

11.1 Introduction

Piccinini (2015b, p. 11ff.) proposes a novel account of when a physical system implements a computation. His idea is that a physical system[1] implements a computation, when the latter mimics the mechanistic properties of the former. In particular, a concrete computing system is a system that has a specific functional mechanism. If such functional mechanism corresponds to a given computation, then we say that the system is implementing such computation. In short, Piccinini's account explains the computational properties of a system through its mechanistic (functional) properties.

Even though Piccinini's idea of connecting the computational properties of a system with its mechanistic properties is interesting, we believe that the way he presents and justifies his account is too abstract, i.e. he does not specify how we could obtain a mechanistic description of a physical system. Our account fills this gap: we specify an account which is similar in spirit with Piccinini's account, expanding it with references to how we could obtain a formal description of a physical system's functional mechanisms. In the paper, we will first introduce the idea of a mapping accounts and explain their core structure and some established difficulties those accounts face (part one); we will then introduce and explain some prerequisite and necessary notions needed by our account (part two); we will then present explicitly our account (part three); we will finally check our proposal with respect to both Piccinini's desiderata (a list of *musts* every account of implementation ought to respect) and classical objections to accounts of implementation (part four). This should be sufficient to show that our account is both in line with Piccinini's ideas (and improving them) and it avoids problems specific of mapping accounts of implementation. Before moving on, we should point out that a by-product of our account is that we introduce a concept of realization[2] which could be useful to pose metaphysical and empirically well-founded questions about concrete computational systems. This is due to our assumption that the fundamental feature of a definition of realization should be that of providing a suitable way to connect the mathematical notion of computation to an actual part of the physical world.

---

[1] We intend the term "a physical system" as a sort of rigid designator, that is a physical system is a certain system without considering any of its characteristic. The theory through which we represent and explain the system sheds light on its peculiarities, but this theory is something that comes after a mere designation of the system. Therefore "physical systems" are not necessarily physical in the sense of physics.

[2] We rather talk about *realization* than *implementation* to make it clearer that the way we are dealing with the problem of computation in physical systems is different from the way computer scientists deal with the same problem. Our goal is to provide a philosophical analysis of the problem and changing the terminology might help in highlighting this gap of approach.

11.2 Mapping Accounts

Mapping accounts of realization rely on the mathematical notion of mapping function. According to mapping accounts of realization, for a physical system to realize a computation there must exist a mapping function between a suitable description of the physical system and a computation. Different mapping accounts differ by which properties the mapping function ought to possess and which descriptions should be employed for the physical systems (e.g. differential equations) and the computations (e.g. Turing Machines. See Turing (1936)).

Piccinini (2015b, pp. 17–18). notes that a first difficulty mapping accounts of realization meet is that (normally) the description of physical systems ascribe to those system an uncountable number of physical states, whereas the description of computations identify countably many computational states. Therefore, there is a gap in the number of states which have to be mapped and providing a suitable mapping function is, in fact, impossible (there aren't enough computational states to which physical states can be mapped). Piccinini also offers a possible answer to this problem: it is possible to tackle this problem reversing the direction of the mapping, i.e. defining it as a function from computational states to physical states (this avoid the problem, since it is always possible to find a physical state for every computational state and this a mapping function can be given). We will indeed follow his suggestion in our proposal.

Piccinini (2015b, p. 19ff.) also discusses at length about the possibility of restricting the mapping function through the introduction of a counterfactual condition, as emphasized by Maudlin (1989), Copeland (1996) and Chalmers (1996). He does so because he, justly, argues that unconditioned mapping functions and the derived unconditioned mapping accounts are too liberal, allowing the possibility of ascribing infinitely many different computations to physical systems. Allowing this possibility is problematic on two (closely connected) grounds: first, it conflates with our scientific understanding of the behaviour of certain physical systems, for which only a limited amount of functions are performed by each physical system and not infinitely many; second, such accounts fail in their explanatory role, since they do not provide a clear explanation of which are the underlying mechanisms of the physical system. Again, we will follow Piccinini's insights and provide a definition of realization in which counterfactuals are justified by scientific laws: we will call (as Piccinini (2015) does) the account based on such a notion "nomological mapping account". The starting point for our definition will be Fano et al. (2016).

Here and in the rest of the paper, we intend laws in a generic non-regularistic way. Confirmed laws establish the causal structure of the system they govern. Even if the causal structure of a system is the ontological grounding of the laws ruling it, the latter are the best description of the former.

Piccinini (2015b, *ibidem*) emphasizes that causal, dispositional and nomological mapping approaches are essentially equivalent. We agree, with the caveat that the latter one is the most parsimonious from an ontological point of view and the nearest to the practice of empirical sciences.

Even though Piccinini admits that counterfactually augmented mapping accounts are superior to their unconditioned counterparts, he still moves some objection to those. What he argues is that these perspectives are:

1. not sufficiently objective, i.e. the counterfactual conditions imposed are often chosen ad hoc and, therefore, the resulting mapping accounts ascribe computations to physical systems according to the modelers insights and not in a objective way;
2. not sufficiently restrictive, i.e. those accounts still ascribe too many computations to physical systems, even ones that are intuitively not ascribable;
3. strongly dependent on the languages employed to describe computations and physical reality, i.e. the choice of computational language (e.g. Turing Machines) or physical language (e.g. differential equations) used to describe the two systems heavily impacts the ascribable computations.

Concerning 1. we argue that in our account objectivity must depend on our best scientific theories, therefore this point is not an issue for our scientifically grounded mapping account. Concerning 2., note that our aims, as pointed out in footnote 1, are more about metaphysical/philosophical questions rather than about computer and cognitive sciences' problems; it is not a coincidence that philosophers with some metaphysical interests, as Chalmers and Maudlin, follow a similar perspective. It will become evident that our approach avoids unlimited pancomputationalism (see discussion in part four), but it is consistent with the possibility that every physical system realizes a computation, therefore avoiding being too restrictive.

Concerning 3., we emphasize that even if we do not accept completely the panlinguistic perspective, which dominated philosophy from the sixties to the nineties, it is difficult to deny that each relevant philosophical concept – included the one of realization – must be anchored to our scientific theories and therefore to the languages in which they are formulated.

Before beginning our investigation, some words on mechanism are in order. In their seminal paper Machamer et al. (2000) established the philosophical agenda of this new perspective. In his book, Piccinini (2015a, b) proposed a mechanistic interpretation of our problem, different both from classical mapping accounts. In this paper, we do not criticise Piccinini's attempt directly, but we develop better the mapping account strategy. As emphasized by Craver and Tabery (2015) a mechanism is characterized by a set of *parts organized* by *causal* connections and producing a *phenomenon*. During our discussion, we will see that this definition, though quite standard, is sufficiently vague that our account could be labelled as a mechanistic account. Indeed, 1. the phenomenon we intend to explain is computation; 2. we do not intend causality in a Humean framework, but we will commit ontologically the least possible; 3. the organization is not something emerging from the system, but is determined by the rules of the computation; 4. parts are individuated by our realization function C. Therefore, in a certain sense, our perspective is mechanistic in spirit, by providing the means to establish the mechanistic behaviour of physical systems, through the use of computational systems and their description.

11.3 Physical Systems as Models of a Theory

To introduce our definition of realization (implementation)[3] we will make use of a simple example. The way the example is constructed makes it possible for us to introduce all the prerequisite notions that are necessary to fully present our definition of realization. The example will be that of a pool table on which different balls are placed randomly. We will use a mathematical abstraction to describe such table and all the balls on it.

Let's call b_i the pool table with all the balls at time i. b_i is nothing more than a proper name for the table and all the balls it contains at time i and, similarly to ordinary proper names in Kripke's interpretation, it refers without describing or making explicit any characteristic or property of the table and the balls. Therefore $\{b_i\}$ is a denumerable set of proper names. It will also be assumed, for this example, that the table is almost completely isolated, i.e. it does not exchange matter or energy with its environment.[4] We will also assume a classical physics context, where space is represented in Euclidean terms and the laws are those of classical mechanics. We will denote with R_{mc} the *state space* used to represent some measurable characteristics[5] (as positions and velocities) of the balls on the table system in time. P is a function associating a point of R_{mc} to each b_i: $P(b_i) = p_i$. Each p_i is a state of the table at time i. We will assume that time is discrete,[6] distinguishable from space and equipped with a linear order: these assumptions stems from the fact that we will be dealing with digital and not analogue computability and because we will analyse situations where relativistic effects are negligible. Given our assumption, the function P brings, at every moment i, b_i in one, and only one, state p_i.

Moving from our framework,[7] it is possible to construct a *transition function F* that takes as argument a couple formed by a point of R_{mc} and a given lapse of time and returns as value another point of R_{mc}, that refers to the state the system is in after the given lapse of time is passed. For example, if p_i is the state of the table corresponding to the name b_i, p_j is the state corresponding to the name b_j and $j>i$, then $P(b_j) = p_j = F(p_i, (j-i))$.

---

[3] From now on we will omit the term "implementation".

[4] The reason for this choice is merely practical. When one evaluates a physical system, it is nearly impossible to be completely rigorous, specifying and keeping track of every possible interaction the system has with its environment and the effects such interactions bring with them. We therefore make use of a good practise from modern natural sciences, bringing the object of our investigation to a partially ideal situation that can find approximate concrete examples in the world. For the purpose of this paper, for example, it is true that, in the real world, physical systems might never be screened from gravitational effects. On the other hand, it is not hard to find situations in which gravity has a physical effect so small that is negligible, e.g. on the computations performed by a personal computer.

[5] Note that those characteristics could also be non-observable.

[6] We will explain in the following how to make time discrete.

[7] It should be remembered that classical mechanics is in principle deterministic, which is fundamental for the construction of the transition function.

More in general we say that the pool table, so constructed, is a *model* of a theory *T* with laws *L* (the laws of classical mechanics) and that discrete time is produced by an *intermediation function I* which ascribes successive instants of time to physical states characterized by a discontinuity. In saying that a certain physical system (the pool table of our example) is a *model* of a theory *T* with laws *L* we are following Beggs and Tucker (2007) terminology. Furthermore, with *intermediation function* we are referring to a function that can transform a continuous state space into a discrete one. The label "intermediation" comes from the fact that the function is used by us to intermediate between physical systems that can be in a continuous set of states and computational states, that are discrete.

What we will claim is that, if a certain physical system is a model of a certain theory *T* (in particular, the best theory at our disposal able to describe such system) with deterministic laws *L* and it is possible to apply our intermediation function to such model, then, if we can construct a *mapping* between all the computational states that form the computation and the discrete state space of the physical system, then that particular physical system realizes the computation.

Perhaps a more realistic example can explain a little more the important role played by the mapping. It is well known that modern digital computers are based on MOSFET technology, also called CMOS. There are two kinds of MOSFETS the N and the P. In the former current pass' only if voltage is high, whereas in the latter when it is low. MOSFETS (due to their size) are toy models of simple quantum mechanics, where relativistic phenomena are not relevant. Since MOSFET are too big to show coherence phenomena, the evolution laws which govern their behaviour are almost deterministic.

The fundamental gate of CMOS technology is the NOR gate, that is the logic gate showing the truth table of "neither... nor...". It is possible to realize a NOR gate building the following simple circuit (Fig. 11.1):

Fig. 11.1 CMOS NOR gate

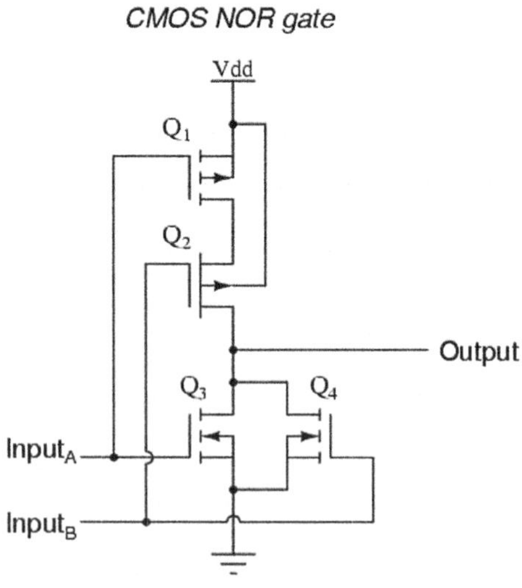

where Q_1, Q_2 and Q_4 are N MOSFET whereas Q_3 is a P MOSFET. It is easy to show that only if the current on $input_A$ and $input_B$ is low on the output the current will have high voltage. Normally to the low voltage corresponds 0 and to the high voltage corresponds 1.

Nevertheless, note that if we go against the common interpretation, that is, if we associate to the low voltage the truth value 1 and to the high voltage the truth value O, the circuit will no longer realize the NOR gate, but the OR gate!

The example shows on one side the central role of a scientific theory – non-relativistic quantum mechanics – in understanding what is a realization; on the other that a correspondence mapping is essential to establish which computation a certain physical system is realizing.[8] In the next section, we will analyse better this mapping.

11.4 Realization

For this paper, we will work with one particular model of computation: Turing machines. The choice is purely arbitrary and nothing of our argument relies on this choice; the argument would still go through if we used Gödel's equational calculus or Church's λ-definability. The *organization* one is looking for is couched in the language of computation one choices. The same would be valid, if we choose a more realistic notion of computation, as register machines. We, nonetheless, prefer to use Turing machines because they end up being more intuitively appealing.

We represent the state[9] of a Turing machine, using Emil Post's elegant formalism (Post 1947), as *instantaneous descriptions* of the form Ps_kq_lQ, where P and Q are variables on (possibly empty) strings of symbols s (in the machines' alphabet) printed on the corresponding cells of the tape, while s_kq_l (where k and l vary over a bounded domain) is the cell in which you will find the machine head, that is the internal state q_l of the head is over a particular cell of the tape and the symbol performed over the scanned cell is s_k. Given the way a Turing machine is constructed (with the tape being potentially infinite) and given the way Ps_kq_lQ represents the state of the machine, it follows that Ps_kq_lQ can be potentially infinite, i.e. as long as you want, but always actually finite (unbounded). We call the state space containing all states R_{TM}. We can think of each Ps_kq_lQ as a point (we will call this point m_i) in R_{TM}.

It is important that the reader notes, before we provide our definition of realization, that computer science, differently from physics (which moves bottom-up, i.e.

[8] Even if the difference between "or" and "nor" is not so essential, since the latter is the dual of the former.

[9] We are improperly using here the term "state" (to be distinguished from the term "internal state") because we wish to highlight the connection between the concepts of computational states of a Turing machine and of physical states. Our reasoning can easily be reformulated using the more familiar term "configuration".

it seeks mathematical representations of part of the world), has a top-down[10] approach, i.e. it seeks implementation of computation in parts of the world. In this respect the notion of realization acquires its importance. It is hard to understand the task of computer scientists that want to find which computations some physical systems perform, if we do not have a clear way of recognizing which physical systems are computing systems.

In fact, we look at the world through the glasses of our best scientific theories; therefore, when we attempt to understand if a physical system is realizing a computation we do not establish a correspondence between points of R_{TM} and elements of the class B of b_i[11], but between R_{TM} and R_{TL} (what in our pool table example is R_{mc}), where R_{TL} is the state space of a physical theory T with laws L. The essential role of T and L guarantees that desideratum 3. is satisfied, that is our framework is scientifically grounded.

However, a Turing machine must not be defined solely in terms of its states, but also by a sufficiently large[12] set of quadruples of the type $s_k q_l O q_m$, where $s_k q_l$ is the internal state of the head and the symbol of the cell in which it is located, while Oq_m indicates that the successive internal state of the head is q_m and O is the operation of head movement either forward or backward or the change of the symbol printed in the head position, or stop.

Therefore, for an *isolated* system w to *realize*[13] a computation described by a Turing machine *TM*, with respect to an *Intermediation function I*, it must hold that:

1. w is a model of a physical theory T with deterministic laws L which apply in w and whose state space is R_{TL}. An *Intermediation function I* must be applied to R_{TL} to make it discrete and to allow a correspondence between the physical system and the computations[14]. At each instant of time, w must be represented by a single point in R_{TL}.
2. There is an injective[15] function C that associates each possible state of *TM*, i.e. arrays of type $Ps_k q_l Q$ (or points m_i of R_{TM}), to a point p_i of R_{TL}, such that if $C(m_j)=p_j$ and $C(m_z)=p_z$, then, if the quadruples of *TM* stipulate that after n steps

---

[10] Something very similar is maintained in Horsman et al. (2014). The latter's approach is comparable to ours in the sense that it emphasizes the importance of the relation between computation and physical space. They express something very similar to what we will call condition 2 of our definition. Despite this, our feeling is that their approach, though interesting, is naïve when dealing with epistemological matters.

[11] Remember that b_is are names that refer to physical states without any intervention of abstraction or modelling. They are somewhat indicators of physical systems as they appear at very specific times.

[12] It is not necessary that there are different quadruples for each couple $s_k q_l$.

[13] We think that it is advisable to start the reflection on the notion of realizability from Giunti's (1997; especially par. 16) important study. Our approach is partially similar from a formal point of view, but it is conceptually different, since it involves the laws of physics.

[14] We know that measured time is discrete as well, since from a technological point of view there is always a minimal threshold of observability.

[15] In general, C will not be surjective, since not all the R_{TL} space is used. Being C injective, the function will be invertible.

m_j goes into m_z, then it must be that – in condition of normal operation[16] – $L(p_j, n)=p_z$.[17]

A few informal considerations about this definition are in order.

First, condition 1. requires us to look at the world through the glasses of our best physical theories and not at the world "directly", that is through our prejudices. Moreover, it highlights the structure of the specific theory, which is adequate to analyse the specific parts of the world we would like to look at.

Second, condition 2. requires that if *TM* leads from a certain input to a certain output, then *C* must be built in such a way that the physical state corresponding to the input must cause deterministically the physical state corresponding to the output in accordance with the *L* laws. Note that our definition is not based, as usual, on an interpretation function of physical states, but on an implementation function from machine states to physical states. This will become important in the third part, when we will compare our definition with what is usually called the *simple mapping account*.

Third, before establishing *C* as required by the definition, it is necessary to make the state space R_{TL} discrete. This is the function of *I* (the intermediation function).[18] In addition, *C* must be so constituted that it does not associate *one by one* a different physical state to each state of the *TM*, since in this case we had to find an infinite series of discrete points in R_{TL}. Therefore, *C* must be based on a code to transform this actual infinite sequence into a finite number of correspondences, which can generate a potentially infinite sequence (like ciphers for numbers, Stabler (1987)). Here we see as *C* establishes the *parts* of the system relevant for the phenomenon "computation".[19]

Fourth, Piccinini (2015b, p. 24) says that a nomological mapping account is not able to take in consideration a miscomputation. On the contrary, in our epistemic approach the theory *T* of which *w* is a model is incomplete in the general meaning of the term. That is *T* is not able to represent all that happens inside *w*. Therefore, in *w* it could occur processes which brings it to a miscomputation. It follows that also our desideratum 2. is satisfied. That is apparent miscomputation is caused by our incomplete knowledge.

The definition we provide can be understood through a simple image, taken from Horsman et al. (2014)'s concept of commuting diagram.

The idea is that if we apply C^{-1} to the state p_i, obtaining m_i, and then perform the computation according to the Turing machine's algorithm, getting to m_z, we get to the same result we would have got to by applying the law *L* to p_i, obtaining p_z, and then applying the C^{-1}, getting, as said, to m_z.

---

[16] This condition will be explained later.

[17] In this case *L* is what we have called in the first part a transition function. This transition function is given by our representation of the laws of nature.

[18] See Scheutz (1999). Discretization is based on a correspondence with discontinuity points in the state space of the physical system.

[19] We do understand that this point is problematic in our approach, especially if we do not specify C explicitly for a given system. We will explore this problem further in future works (Fig. 11.2).

Fig. 11.2 Commuting diagram

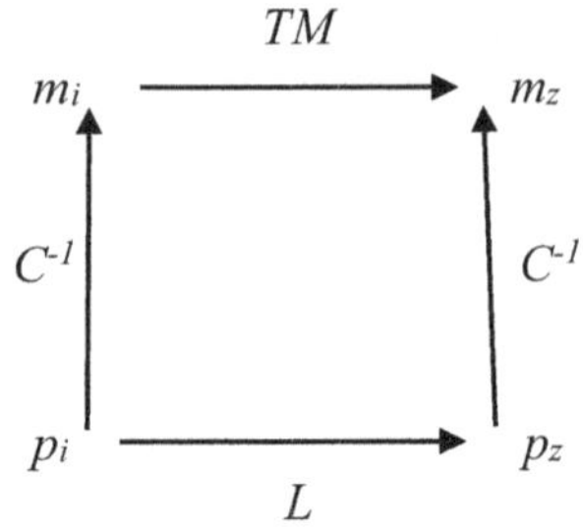

11.5 Assessing the Definition

To assess the account based on the definition we have just provided, we will employ some of Piccinini's (2015b, p. 11ff) desiderata. In his book, Piccinini introduces six desiderata that every good account of realization ought to fulfil. The desiderata are, in order:

1. *Objectivity*: the computations ascribed must be objectively identifiable and they should not depend, for their ascription, to subjective interpretations;
2. *Explanation*: the computations ascribed should provide some insight into the behaviour of the physical system;
3. *The right things compute*: paradigmatic examples of computing physical systems (e.g. computers) should be ascribed, by the account, the right computations;
4. *The wrong things don't compute*: paradigmatic examples of non-computing physical systems (e.g. rocks) shouldn't be ascribed any computations;
5. *Miscomputation is explained*: examples of miscomputations should be accounted for in the accounts;
6. *Taxonomy*: the account should be able to explain the different complexities of computational devices.

Even though those desiderata bring up many interesting aspects, we believe that they are not all necessary to assess a philosophical notion of realization. In fact, we believe that desiderata (2), (3), (4) and (6) are too specific and should be taken into consideration only for accounts employed in computer or cognitive sciences' researches. We will then substitute those four desiderata with a broader one, for which a good mapping account should neither be too liberal nor too restrictive; that is, neither unlimited pancomputationalism (the view for which every physical system realizes every computation) nor the view that some systems do not perform any computation should be allowed. Even though this desideratum seems equal to Piccinini's, the main difference is that we won't refer to paradigmatic systems, leaving intuitions about systems specific of different disciples out of the matter (e.g. a brain might be a paradigmatic computational device for cognitive scientists and it might not be for psychologists).

Before moving on, we should clarify better what we mean with our desideratum.

There are many forms of pancomputationalism; Piccinini (2015b, p. 54ff.) argues that even a *limited* form of it is unacceptable, since it is against scientific practice. "Limited" means that *each* physical system realizes a limited quantity of computations, sometimes only one. We believe that this isn't completely true. The main point is that not all objects perform computations at the same level of description; however, when we allow for different levels of description (e.g. macro and micro physical levels of descriptions), we believe that something which seems not to compute, *can*[20] contain parts that do actually compute, transferring their computational nature to the objects they are part of. That is not only laptops and brains, but also DNA, quantum particles, etc. could realize computations and, therefore, the objects they are part of.

We now move to the assessment of the account. We will see that when answering classical objections to mapping accounts, we will also fulfil the posed desiderata.

Concerning objectivity, it must be said that deciding whether a certain physical system either realizes or not a certain computation is a question which depends essentially on the theory we are using to explain the behaviour of the physical system. Therefore, a good definition (and especially ours) could be objective only as far as our best theories are objective. Moreover, our definition, as suggested by Cotogno (2003, pp. 187–188), specifies the notion of 'realization', making explicit the fact that it is not a relationship between the mathematical concept of a Turing machine and a part of the world, but between the former and the physical theory partially true for that part of the world. This is an important upgrade[21] with respect to preceding literature on the subject and avoids difficulties such as: how is it possible for a concrete physical system to implement an abstract object? We avoid the difficulties by saying that all we can know is that the *model* we give according to the theory T and laws L realizes a computation, without saying anything directly about the physical system itself. Thus, in this case, TL will be a physical theory endowed with various principles.

Second, our definition is not based, as usual, on an interpretation function from the states (or subset of states) ascribed to the physical system by a microphysical description to the states defined by a computation description, but, rather, on an implementation function from machine states to physical states. This has, as an advantage, that the subset of physical states to which the computational ones are mapped, must not be determined in advance, but can be fixed according to the computation we want to check.[22] Moreover, we avoid the problem of the difference in cardinality between the two sets of physical states and computation states, since our mapping function is evaluated on an unbounded discrete finite set and has an unbounded discrete and finite range to map onto. This avoids the difficulty of having to map univocally a non-denumerable infinite amount of values onto an unbounded discrete codomain. In this sense, we stand in a better position when compared to *traditional mapping accounts*.

[20] Considering this characteristic we could label our view as *modal limited pancomputationalism*.

[21] See also Horsman et al. (2014).

[22] This opens the door to the possibility of pancomputationalism. Soon we face this point.

Third, our definition is sufficiently restrictive to block Putnam's (1988, pp. 121–125) theorem, according to which each open physical system would be able to realize every automatic procedure.[23] This is because our system is isolated, so that in our case what Putnam calls the Principle of Noncyclical Behaviour, i.e. a "system is in different maximal states at different times" (Putnam 1988, p. 121) does not hold. The principle, fundamental for Putnam's proof, "will hold true for all systems that can 'see' (are not shielded from electromagnetic and gravitational signals from) a clock. Since there are natural clocks from which no ordinary open system is shielded, all such systems satisfy the principle" (Putnam 1988, p. 121). But our system is isolated and so cannot see any external clock. Our definition avoids Putnam's objection also because it is based on the capacity of natural laws to justify counterfactuals. Putnam's theorem, in fact, is founded on a mere *a posteriori* projection of a computation on a physical system; in our perspective, on the contrary, w can realize a computation with different inputs and outputs. Therefore, our desideratum 2. is satisfied as well, since pancomputationalism is avoided.

Finally, our definition avoids Kripke's objection (Kripke, 2013), discussed by Stabler (1987) and Scheutz (2012), according to which the respect of laws L is not a sufficient clause to distinguish a functioning machine from a broken one, because we have added the further clause "in condition of normal operation". The introduction of this surreptitiously normative assumption is due to the fact that, in our opinion, there is no principle able to respond to Kripke's objection. It is necessary to adopt a *design stance* with respect to the part of the world that we are considering, to overcome Kripke's objection. The endorsed design could, nonetheless, be changed. Therefore, the proposed solution can only be a form or reflective equilibrium between the normative and the explanatory point of view.

Those simple answers to standard objections should show how we deal with the desiderata. We comply with the objectivity desideratum by stating that as far as scientific laws are objective, our account is objective, since the latter is based on the former. We comply with the scope desideratum (the account being neither too liberal, not too restrictive) by avoiding unlimited pancomputationalism, but allowing a limited form of it (see Putnam's objection). Finally, we comply with the miscomputation desideratum by assuming a design stance when it comes to Kripke's objection.

[23] Using the words of Chalmers: "The ambitions of artificial intelligence rest on a related claim of computational sufficiency" (Chalmers 1996, p. 309), that is, "the right kind of computational structure suffices for the possession of a mind" (Chalmers 2011, p. 325). So, "computation will provide a powerful formalism for the replication and explanation of mentality" (Chalmers 1996, pp. 309–310). Hilary Putnam's theorem says that "every ordinary open system is a realization of every abstract finite automaton" (Putnam 1988, p. 121), and its proof requires two physical principles; a principle of continuity, and the principle of Noncyclical behaviour (for more details see Putnam 1988, pp. 120–125). "Together with the thesis of computational sufficiency, this [theorem] would imply that a rock has a mind." [...] "We must either embrace an extreme form of panpsychism or reject the principle on which the hopes of artificial intelligence rest. Putnam himself takes the result to show that computational functionalism cannot provide a foundation for a theory of mind" (Chalmers 1996, pp. 309–310).

11.6 Concluding Remarks

Accounts of what means that a physical system realizes a computation could have at least two different aims: either to give a pragmatical definition of what is a computational system, or a precise definition of what it means that a certain physical system implements a given computation. In this paper, we faced the latter problem. We looked for an adequate formalism able to put in correspondence the states of a Turing machine and those of the physical system. For this reason, we built the mapping C between the two states spaces. Our approach does not allow every possible correspondence, but only those supported by scientific laws.

We propose an approach quite different from the mechanistic one developed by Piccinini. Despite this, the standard definition of mechanism is so generic that, if one avoids its typical excessive ontologism, even our perspective could be dubbed mechanistic.

Our investigation has many limitations. We do not consider relativity and quantum physics, which surely have an important impact on our framework. Moreover, our point of view is non-constructive, since we do not outline how to build the C function. Furthermore, indeterminism and dissipative phenomena should be examined.

Despite this, we think we have given an important contribution toward a precise characterization of what is a realization. Classical metaphysical questions, e.g. "could human intelligence (produced by our brain) be represented by a universal Turing machine?", could be faced only after a clear definition of what a realization is and our definition does a great job at advancing the understanding of such notion.

References

Beggs, E.J., and J.V. Tucker. 2007. Can Newtonian Systems, Bounded in Space, Time, Mass and Energy, Compute All Functions? *Theoretical Computer Science* 371 (1): 4–19.

Chalmers, D.J. 1996. Does a Rock Implement Every Finite-State Automaton? *Synthese* 108: 309–333.

———. 2011. A Computational Foundation for the Study of Cognition. *Journal of Cognitive Science* 12 (4): 323–357.

Copeland, B.J. 1996. What Is Computation. *Synthese* 108: 224–259.

Cotogno, P. 2003. Hypercomputation and the Physical Church-Turing Thesis. *British Journal for the Philosophy of Science* 54 (2): 181–223.

Craver C., and J. Tabery. 2015. Mechanisms in Science. *The Stanford Encyclopedia of Philosophy*. Edward N. Zalta (ed.). https://plato.stanford.edu/archives/spr2017/entries/science-mechanisms/

Fano, V., P. Graziani, R. Macrelli, and G. Tarozzi. 2016. Are Gandy Machines Really Local? In *Computing and Philosophy*, Synthese Library, ed. Vincent Müller, 27–44. Cham: Springer.

Giunti, M. 1997. *Computation, Dynamics and Cognition*. Oxford: Oxford University Press.

Horsman, C., S. Stepney, R. Wagner, and V. Kendon. 2014. When Does a Physical System Compute? *Proceeding of the Royal Society A* 470. http://rspa.royalsocietypublishing.org/content/royprsa/470/2169/20140182.full.pdf

Kripke, S.A. 2013. The Church-Turing 'Thesis' as a Special Corollary of Gödel's Completeness Theorem. In *Computability: Turing, Gödel, Church, and Beyond*, ed. B.J. Copeland, C.J. Posy, and O. Shagrir, 77–104. Cambridge, MA: MIT Press.
Machamer, P., L. Darden, and C.F. Craver. 2000. Thinking About Mechanisms. *Philosophy of Science* 67: 1–25.
Maudlin, T. 1989. Computation and Consciousness. *Journal of Philosophy* 86: 407–432.
Piccinini, G. 2015a. Computation in Physical Systems. *The Stanford Encyclopedia of Philosophy*, (Summer 2015 Edition), Edward N. Zalta (ed.). http://plato.stanford.edu/archives/sum2015/entries/computation-physicalsystems/
———. 2015b. *Physical Computation: A Mechanistic Account*. Oxford: Oxford University Press.
Post, E. 1947. Recursive Unsolvability of a Problem of Thue. *Journal of Symbolic Logic* 12: 1–11.
Putnam, H. 1988. *Representation and Reality*. Cambridge, MA: MIT Press.
Scheutz, M. 1999. When Physical Systems Realize Functions. *Minds and Machines* 9 (2): 161–196.
———. 2012. What It Is Not to Implement a Computation: A Critical Analysis of Chalmers' Notion of Implementation. *Journal of Cognitive Science* 13 (1): 75–106.
Stabler, E.P., Jr. 1987. Kripke on Functionalism and Automata. *Synthese* 70 (1): 1–22.
Turing, A.M. 1936. On Computable Numbers with an Application to the Entscheidungsproblem. In *Proceedings of the London Mathematical Society* (42(2), 230–265). A correction in (43(2), 544–546, 1937). Oxford: Oxford University Press.

The manufacturer's authorised representative in the EU is Springer Nature Customer Service Centre GmbH, Europaplatz 3, 69115 Heidelberg, Germany. If you have any concerns regarding our products, please contact ProductSafety@springernature.com

Printed and bound by CPI Group (UK) Ltd, Croydon, CR0 4YY
30/09/2026
02225444-0008